Mosasaurus

-240 -220 -200 -180 -160 -140 -120 -100	millions of years ago

-201.3   -145   -66

Triassic   Jurassic   Cretaceous

Time Period

Mosasaurus was a carnivorous marine reptile that inhabited North America, Europe and Japan. Its name means "Lizard of the Meuse River." It was named after the river in France where the first fossilized specimen was found. It is supposed that the body was very sturdy and that it preyed on various creatures in the sea. Recent research proposes that Mosasaurus had a large crescent-shaped caudal (or tail) fin.

- About 40-60 feet or 12-18 meters long and weighed around 10 tons.

KUM⊙N

Grade **4-5**

45 Days of Kumon

How to use this sheet

Date   /   /

Math   **Reading**

Paste a sticker here for completing a math exercise.

Paste a sticker here for completing a reading exercise.

Try this fitness exercise after finishing math and reading exercises.

Go for a walk.

1

Keep your skills sharp all summer long!

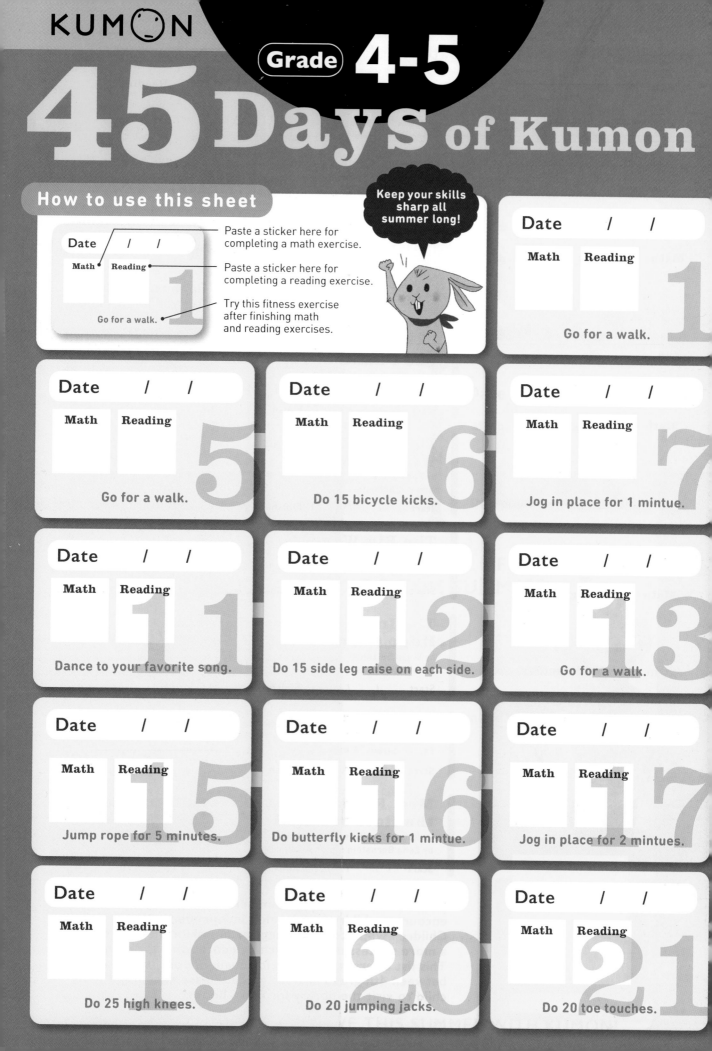

Date   /   /

Math   **Reading**

Go for a walk.

1

Date   /   /

Math   **Reading**

Go for a walk.

5

Date   /   /

Math   **Reading**

Do 15 bicycle kicks.

6

Date   /   /

Math   **Reading**

Jog in place for 1 mintue.

7

Date   /   /

Math   **Reading**

Dance to your favorite song.

11

Date   /   /

Math   **Reading**

Do 15 side leg raise on each side.

12

Date   /   /

Math   **Reading**

Go for a walk.

13

Date   /   /

Math   **Reading**

Jump rope for 5 minutes.

15

Date   /   /

Math   **Reading**

Do butterfly kicks for 1 mintue.

16

Date   /   /

Math   **Reading**

Jog in place for 2 mintues.

17

Date   /   /

Math   **Reading**

Do 25 high knees.

19

Date   /   /

Math   **Reading**

Do 20 jumping jacks.

20

Date   /   /

Math   **Reading**

Do 20 toe touches.

21

After finishing each day's math and reading exercises, paste a sticker
You can also try to complete each day's fitness challenge after your
and complete the Summer Bucket List below. Stay active and learn

To parents: If your child seems to have difficulty pasting the stickers
need help completing the fitness challenges, if your child needs help
Encourage your child to use the fitness challenges included in the pr
60 minutes of daily physical activity.
When he or she has completed all of the pages, offer lots of praise a
Finally, please fill out your child's name and sign your name in the b   2 mintues.

Date   /   /

Math	Reading

Do 25 hops on each foot.

Date   /   /

Math	Reading

Do 15 jumping jacks.

Date   /   /

Math	Reading

Do 15 toe tou

Date   /   /

Math	Reading

Do 10 crisscross jumps.

Date   /   /

Math	Reading

Do 5 sit-u

Date   /   /

Math	Reading

Do 20 high knees.

Date   /   /

Math	Reading

Go for a walk.

Date   /   /

Math	Reading

Do 20 sit ups.

list is designed to include a variety of genres,
ng styles, cultures and authors. Please use
ist as a guide for reading during the summer.

to use this list:
e the date you start / finish reading the book.
to read all 6 books. After you finish a book
can rate it using the star chart below.

om Tollbooth
lustrated
dom House
   ★ ★ ★ ★ ★

.   Finish   /   /   .

Who Loved
es
   ★ ★ ★ ★ ★

.   Finish   /   /   .

ave
   ★ ★ ★ ★ ★

.   Finish   /   /   .

Summer Bucket List

CR
TO

tion of
et
   ★ ★ ★ ★ ★

● Build a tower
with a materia

.   Finish   /   /   .

●
....

●
....

lasting
   ★ ★ ★ ★ ★

● Play a board g

●
....

roux

●
....

.   Finish   /   /   .

● Plant a vegetab

●
....

by and
f NIMH
n, Illustrated
Simon & Schuster
   ★ ★ ★ ★ ★

●
....

● Play act a scene

.   Finish   /   /   .

●
....

●
....

Kumon Summer Reading List is designed to
en to develop independent reading skills.
quire strong reading skills often enjoy
al and enriching educational experience.
uggestions for quality books for readers
de and 5th grade. Please encourage your
library for more books.

● Draw plans for

●
....

●
....

STAY ACTIV

Multiplication

Date   /   /

Name

1 Multiply.

(1)
$$\begin{array}{r} 1\ 3 \\ \times\quad 2 \\ \hline \end{array}$$

(6)
$$\begin{array}{r} 4\ 3 \\ \times\quad 4 \\ \hline \end{array}$$

(11)
$$\begin{array}{r} 3\ 1 \\ \times\quad 7 \\ \hline \end{array}$$

(16)
$$\begin{array}{r} 5\ 7 \\ \times\quad 9 \\ \hline \end{array}$$

(2)
$$\begin{array}{r} 3\ 4 \\ \times\quad 2 \\ \hline \end{array}$$

(7)
$$\begin{array}{r} 2\ 4 \\ \times\quad 5 \\ \hline \end{array}$$

(12)
$$\begin{array}{r} 5\ 3 \\ \times\quad 7 \\ \hline \end{array}$$

(17)
$$\begin{array}{r} 6\ 0 \\ \times\quad 4 \\ \hline \end{array}$$

(3)
$$\begin{array}{r} 2\ 3 \\ \times\quad 3 \\ \hline \end{array}$$

(8)
$$\begin{array}{r} 3\ 7 \\ \times\quad 5 \\ \hline \end{array}$$

(13)
$$\begin{array}{r} 2\ 5 \\ \times\quad 8 \\ \hline \end{array}$$

(18)
$$\begin{array}{r} 7\ 3 \\ \times\quad 2 \\ \hline \end{array}$$

(4)
$$\begin{array}{r} 4\ 3 \\ \times\quad 3 \\ \hline \end{array}$$

(9)
$$\begin{array}{r} 2\ 7 \\ \times\quad 6 \\ \hline \end{array}$$

(14)
$$\begin{array}{r} 3\ 9 \\ \times\quad 8 \\ \hline \end{array}$$

(19)
$$\begin{array}{r} 3\ 6 \\ \times\quad 3 \\ \hline \end{array}$$

(5)
$$\begin{array}{r} 3\ 2 \\ \times\quad 4 \\ \hline \end{array}$$

(10)
$$\begin{array}{r} 4\ 8 \\ \times\quad 6 \\ \hline \end{array}$$

(15)
$$\begin{array}{r} 1\ 6 \\ \times\quad 9 \\ \hline \end{array}$$

(20)
$$\begin{array}{r} 8\ 4 \\ \times\quad 5 \\ \hline \end{array}$$

Vocabulary
Syllables

Date / /

Name

① Read each word aloud. Then divide the word into syllables.

10 points per qu

(1) airplane _____air/plane_____

(2) parachute _____

(3) scrapbook _____

(4) automobile _____

(5) highway _____

(6) computer _____

(7) basketball _____

(8) spectators _____

② Write the words with the same amount of syllables in each group below.

20 points for co

alligator   propeller   hollow   overboard   pelican
motorcycle   dictionary   badger   hurdle

(a) 2 syllables   (b) 3 syllables   (c) 4 syllables

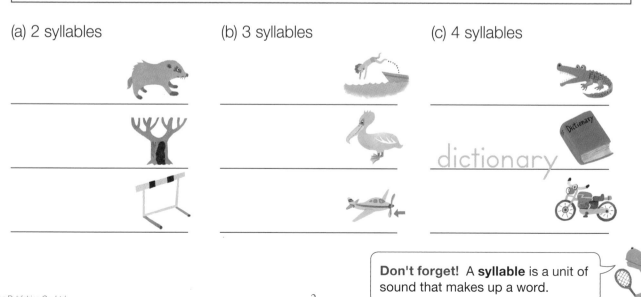

_____   _____   _____

_____   _____   _____dictionary_____

_____   _____   _____

> **Don't forget!** A **syllable** is a unit of sound that makes up a word.

Multiplication

Date   /   /

Name

1 Multiply.

5 points per question

(1)
$$130 \times 4$$

(2)
$$273 \times 2$$

(3)
$$408 \times 3$$

(4)
$$318 \times 3$$

(5)
$$116 \times 4$$

(6)
$$347 \times 4$$

(7)
$$227 \times 5$$

(8)
$$547 \times 5$$

(9)
$$409 \times 6$$

(10)
$$647 \times 6$$

(11)
$$503 \times 7$$

(12)
$$381 \times 7$$

(13)
$$308 \times 8$$

(14)
$$459 \times 8$$

(15)
$$207 \times 9$$

(16)
$$728 \times 9$$

(17)
$$274 \times 6$$

(18)
$$686 \times 7$$

(19)
$$778 \times 8$$

(20)
$$889 \times 9$$

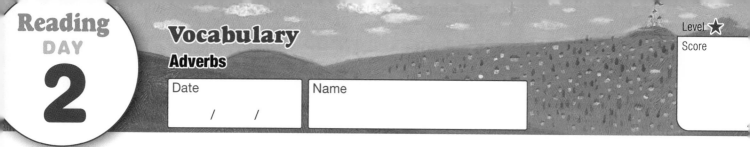

Vocabulary
Adverbs

Level ⭐

Date   /   /

Name

Score

① Complete the table below according to the example.

50 poi
for

adjective	adverb
immediate	immediately
deliberate	
normal	
polite	
rapid	
playful	
swift	

adjective	adverb
lazy	lazily
sleepy	sleepily
merry	
helpful	
generous	
light	
dainty	daintily

② Read the passage. Then answer the questions below using only adverbs from the passage.

10 poin
per (

An old woman with a large bag boarded a bus that rapidly drove away. Immediately, a young man generously gave his seat to her. Normally, most people would swiftly take the seat and politely thank the man for being so helpful. Instead, the woman daintily laid her bag on the seat and remained standing. Was the woman deliberately being rude? No, she merrily explained that inside the bag was a litter of sleepy kittens. She was carefully taking them to a new home.

(1) How did the bus drive?

The bus drove _____.

(2) When did the young man give up his seat?

The young man gave up his seat _____.

(3) How would people normally react to the young man's offer?

Normally, most people would _____ thank the man.

(4) How did the woman lay her bag down?

The woman laid her bag down _____.

(5) How was the woman taking the kittens to a new home?

The woman was _____ taking the kittens to a new home.

Don't forget! An **adverb** is a word tha describes a verb. Adverbs usually hav "ly" as a suffix.

Multiplication

Date   /   /

Name

Score

/100

1 Multiply.

5 points per question

(1)
$$\begin{array}{r} 32 \\ \times\ 12 \\ \hline \end{array}$$

(6)
$$\begin{array}{r} 46 \\ \times\ 42 \\ \hline \end{array}$$

(11)
$$\begin{array}{r} 32 \\ \times\ 74 \\ \hline \end{array}$$

(16)
$$\begin{array}{r} 54 \\ \times\ 91 \\ \hline \end{array}$$

(2)
$$\begin{array}{r} 42 \\ \times\ 23 \\ \hline \end{array}$$

(7)
$$\begin{array}{r} 23 \\ \times\ 52 \\ \hline \end{array}$$

(12)
$$\begin{array}{r} 53 \\ \times\ 72 \\ \hline \end{array}$$

(17)
$$\begin{array}{r} 50 \\ \times\ 41 \\ \hline \end{array}$$

(3)
$$\begin{array}{r} 43 \\ \times\ 34 \\ \hline \end{array}$$

(8)
$$\begin{array}{r} 38 \\ \times\ 55 \\ \hline \end{array}$$

(13)
$$\begin{array}{r} 29 \\ \times\ 81 \\ \hline \end{array}$$

(18)
$$\begin{array}{r} 67 \\ \times\ 24 \\ \hline \end{array}$$

(4)
$$\begin{array}{r} 54 \\ \times\ 37 \\ \hline \end{array}$$

(9)
$$\begin{array}{r} 47 \\ \times\ 60 \\ \hline \end{array}$$

(14)
$$\begin{array}{r} 34 \\ \times\ 83 \\ \hline \end{array}$$

(19)
$$\begin{array}{r} 35 \\ \times\ 38 \\ \hline \end{array}$$

(5)
$$\begin{array}{r} 34 \\ \times\ 45 \\ \hline \end{array}$$

(10)
$$\begin{array}{r} 28 \\ \times\ 61 \\ \hline \end{array}$$

(15)
$$\begin{array}{r} 26 \\ \times\ 93 \\ \hline \end{array}$$

(20)
$$\begin{array}{r} 84 \\ \times\ 58 \\ \hline \end{array}$$

Vocabulary
Silent Letter Words

Date
/   /

Name

Level ⭐
Score

1 Trace the words below.

55 points for com

(1) knead
(2) wrench
(3) kneel
(4) knuckles
(5) bomb
(6) pneumonia
(7) limb
(8) cologne
(9) wreckage

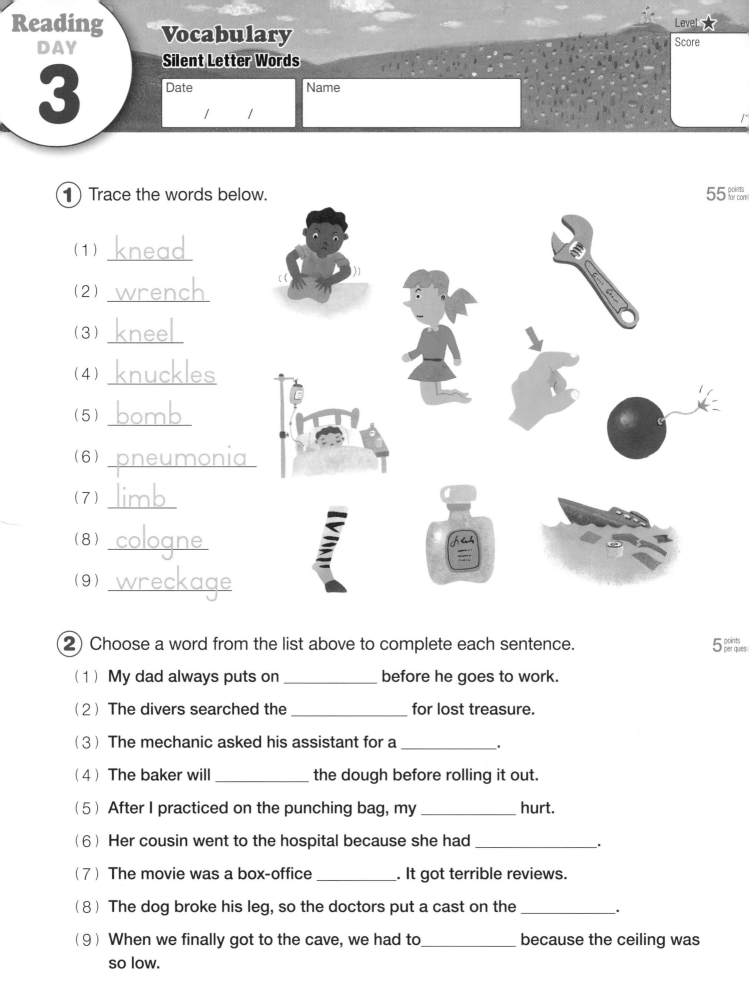

2 Choose a word from the list above to complete each sentence.

5 points per ques

(1) My dad always puts on _____ before he goes to work.

(2) The divers searched the _____ for lost treasure.

(3) The mechanic asked his assistant for a _____.

(4) The baker will _____ the dough before rolling it out.

(5) After I practiced on the punching bag, my _____ hurt.

(6) Her cousin went to the hospital because she had _____.

(7) The movie was a box-office _____. It got terrible reviews.

(8) The dog broke his leg, so the doctors put a cast on the _____.

(9) When we finally got to the cave, we had to_____ because the ceiling was so low.

Multiplication

1 Multiply.

5 points per question

(1)
```
   3 2 2
 ×   1 3
```

(2)
```
   3 2 2
 ×   3 3
```

(3)
```
   3 1 4
 ×   1 4
```

(4)
```
   3 1 4
 ×   4 5
```

(5)
```
   4 0 7
 ×   2 4
```

(6)
```
   4 2 3
 ×   5 1
```

(7)
```
   9 0 6
 ×   3 7
```

(8)
```
   3 1 6
 ×   7 0
```

(9)
```
   6 0 9
 ×   5 5
```

(10)
```
   3 7 0
 ×   3 8
```

(11)
```
   1 3 5
 ×   1 6
```

(12)
```
   5 3 4
 ×   4 7
```

(13)
```
   4 1 2
 ×   6 7
```

(14)
```
   6 1 9
 ×   5 8
```

(15)
```
   2 7 0
 ×   5 0
```

(16)
```
   1 6 4
 ×   5 6
```

(17)
```
   6 7 2
 ×   3 2
```

(18)
```
   6 0 8
 ×   8 9
```

(19)
```
   7 3 1
 ×   4 4
```

(20)
```
   3 4 5
 ×   9 5
```

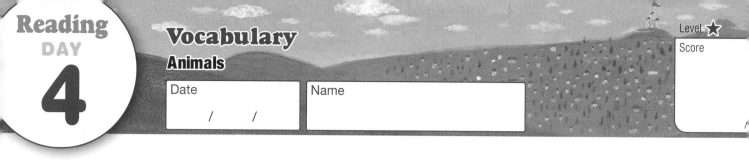

Reading
DAY
4

Vocabulary
Animals

Date	Name
/ /	

Level ★
Score

① Complete the passage using vocabulary words defined below.

10 points
per qu

The sloth is an animal that lives up to its name, which means "laziness." These (1) mammals are (2)_____ for being slow and sleeping up to twenty hours a day. Sloths (3)_____ in the trees of the tropical forests in Central and South America. While their (4)_____ arms and wooly fur make them look like monkeys, they are more closely related to armadillos and anteaters. There are two main (5)_____ of sloth. Sloths with two toes hang upside-down, while sloths with three toes like to sit upright. Three-toed sloths also have an extra (6)_____ in their necks so they can turn their heads almost all the way around. Both types of sloth are slow. In fact, they're so slow that (7)_____ grows on their fur. Some scientists think that sloths move slow so (8)_____ won't see them. The green algae also acts as (9)_____. But they're not only slow moving—a sloth can take up to a month to (10)_____ one meal.

infamous	have a bad reputation
vertebrae	a section of bone or cartilage that make up the spinal column
dwell	to stay for a while; to live in a place
algae	any plant or plantlike living creature similar to seaweed
species	a category of living things; a class of things of the same kind and with the same name
digest	to break down food and absorb it in the body
predators	animals that lives by killing and eating other animals
camouflage	the hiding or disguising of something by covering it up or changing the way it looks
mammals	warm-blooded animals with verterbrae that feed their babies with window.
lengthy	very long

Multiplication

Date   /   /

Name

Score

/100

1 Multiply.

5 points per question

(1)
```
  1140
×    2
```

(6)
```
  1307
×    7
```

(11)
```
   132
× 123
```

(16)
```
   230
× 125
```

(2)
```
  1273
×    3
```

(7)
```
  1005
×    8
```

(12)
```
   115
× 134
```

(17)
```
   216
× 107
```

(3)
```
  2124
×    4
```

(8)
```
  1084
×    9
```

(13)
```
   122
× 146
```

(18)
```
   204
× 109
```

(4)
```
  2315
×    5
```

(9)
```
  2004
×    6
```

(14)
```
   213
× 158
```

(19)
```
   270
× 261
```

(5)
```
  3112
×    6
```

(10)
```
  2107
×    4
```

(15)
```
   201
× 113
```

(20)
```
   163
× 310
```

Vocabulary
Science

Date   /   /

Name

① Complete the passage using vocabulary words defined below.   **10** points per que.

Get a ping-pong ball, a rubber ball, and a wooden ball of the same size. Place all three in water. Ever (1)_____ why the ping-pong ball floats the best and the wooden ball is almost underwater? It's easy to see that light objects filled with air, like

(2)_____ rafts, are good at floating. But solid things can float too. Why? A Greek (3)_____ named Archimedes, who was born around 287 BC, was able to explain this

(4)_____. Archimedes' first (5)_____ was that floating objects are held up by a thrust called

(6)_____. His second idea was that

the force needed to keep the object (7)_____ was equal

to how much water the object (8) _____. After a lot of

(9)_____, Archimedes could (10)_____ that the amount of upward force on a floating object is equal to the weight of the water it moves out of place.

ponder	think about something carefully
inflatable	able to fill with air or gas
phenomenon	a fact, feature, or event of scientific interest
scientist	a person skilled in science
theory	an idea that is the starting point for argument or investigation
experiments	tests; operations carried out in order to discover something
afloat	carried on, or as if on, the water
buoyancy	the tendency to float or to rise when in a fluid
displaced	removed from an usual or proper place
prove	to show the truth by evidence

Multiplication

Level ★★

Score

/100

Math
DAY
6

Date   /   /

Name

1 Multiply.

5 points per question

(1)   6 1
     × 8

(2)   5 7 2
     ×   7

(3)   4 7
     × 5

(4)   6 6
     × 8

(5)   2 4 5
     ×   4

(6)   8 0 9
     ×   7

(7)   3 2 0 4
     ×   5

(8)   6 5 5 2
     ×   9

(9)   1 0 4
     ×   9

(10) 7 5 3
     ×   7

(11) 6 9
     × 2 9

(12) 1 0 5
     × 2 4

(13) 2 4 6
     × 5 3

(14) 9 3
     × 3 9

(15) 7 4 1
     × 5 6

(16) 2 3 0
     × 8 0

(17) 8 7
     × 3 0

(18) 6 1 8
     × 7 3

(19) 5 0
     × 4 6

(20) 2 0 8
     × 3 1 9

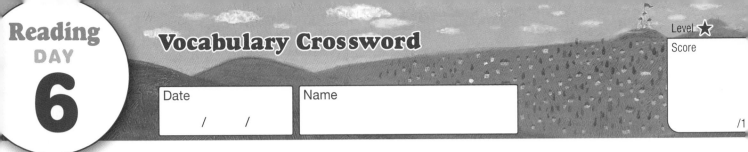

Vocabulary Crossword

Date   /   /

Name

① Complete the crossword puzzle using the sentences below as clues.   **10** points per ques

ACROSS

(1) Groups of monkeys still _____ in these forests.

(2) The stunt woman opened her _____ about 800 meters or 2600 feet from the ground.

(3) The man had sprayed on too much _____.

(4) A new _____ of frog was discovered by scientists.

(5) The man was saved from the _____ of the crashed car.

DOWN

(6) We got on the wrong _____ and drove in the wrong direction.

(7) I _____ finished my test so I rechecked my answers.

(8) The chef _____ crushed pepper over the salad.

(9) Her family _____ in church when they are praying.

(10) Our music teacher tried to explain the _____ behind harmonies.

Division

Date   /   /

Name

Score

/100

1 Divide.

4 points for completion

(1)

$$2 \overline{) 1\ 4}$$

(2)

$$2 \overline{) 2\ 5} \quad \square\square R\square$$

(3)

$$2 \overline{) 3\ 0} \quad \square\square$$

(4)

$$2 \overline{) 5\ 4}$$

(5)

$$3 \overline{) 2\ 4}$$

(6)

$$3 \overline{) 2\ 8} \quad \square R\square$$

(7)

$$3 \overline{) 3\ 6}$$

(8)

$$4 \overline{) 3\ 2}$$

(9)

$$4 \overline{) 4\ 1}$$

(10)

$$4 \overline{) 5\ 6}$$

(11)

$$5 \overline{) 4\ 0}$$

(12)

$$5 \overline{) 4\ 3}$$

(13)

$$5 \overline{) 6\ 0}$$

(14)

$$6 \overline{) 3\ 0}$$

(15)

$$6 \overline{) 5\ 5}$$

(16)

$$6 \overline{) 7\ 8}$$

(17)

$$7 \overline{) 4\ 2}$$

(18)

$$7 \overline{) 5\ 7}$$

(19)

$$7 \overline{) 8\ 4}$$

(20)

$$8 \overline{) 4\ 8}$$

(21)

$$8 \overline{) 6\ 0}$$

(22)

$$8 \overline{) 9\ 0}$$

(23)

$$9 \overline{) 5\ 4}$$

(24)

$$9 \overline{) 6\ 5}$$

(25)

$$9 \overline{) 9\ 9}$$

Keep up the good work!

Reading
DAY
7

Defining Words by Context
Canada

Level ⭐

Score

Date   /   /

Name

/1

1 Read the short passage. Then choose words from the passage to complete the definitions below.

10 points for com

Canada is the second largest country in the world, but it only has half of one percent of the world's **population**. That means a lot of open space. Canada has lakes, rivers, mountains, **plains**, forests, and swamps. It even has the only **temperate** rain forest in the world. Canada **spans** more than half of the Northern **Hemisphere**. In the far north of Canada, you can see ice, snow, and **glaciers**.

With all this space comes many different animals—bears, mountain lions, otters, and many freshwater fish. Canadians **cherish** nature and wildlife. Forty-one national parks and three marine **conservation** areas have been made to protect animals like the wolf and **lynx**. These animals need to be protected because they have been **overhunted**.

(1) _____ large bodies of ice that move slowly

(2) _____ to hold dear; to keep with care and affection

(3) _____ a careful protection of something

(4) _____ broad areas of level or rolling treeless country

(5) _____ a large wild cat

(6) _____ the whole number of people living in a country or region

(7) _____ hunted too much

(8) _____ a climate that is usually mild without very cold or hot temperatures

(9) _____ reaches or extends across

(10) _____ half of the earth

14

Date   /   /

Name

1 Divide.

4 points per question

(1) 2)224

(8) 4)408

(15) 6)726

(22) 9)198

(2) 2)208

(9) □□R□ 4)350

(16) 7)455

(23) 9)354

(3) 2)150

(10) 5)350

(17) 7)721

(24) 9)505

(4) 3)150

(11) 5)570

(18) 7)500

(25) 9)963

(5) 3)315

(12) 5)473

(19) 8)840

(6) □□□R□ 3)320

(13) 6)630

(20) 8)454

Outstanding job!

(7) 4)140

(14) 6)450

(21) 8)616

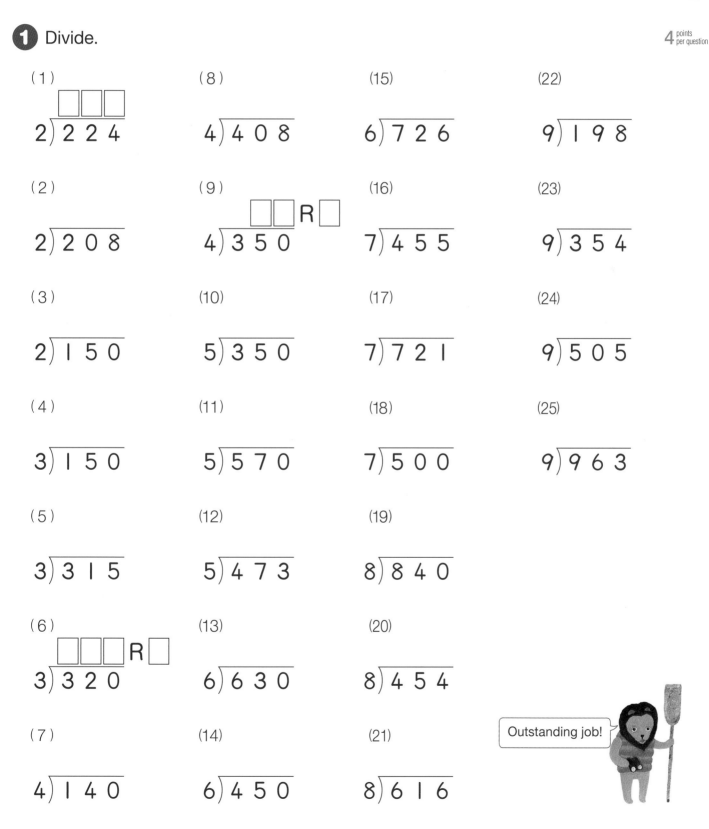

Defining Words by Context
Making Light

Level ★

Score

Date   /   /

Name

① Read the short passage. Then choose words from the passage to complete the definitions below.   **10** points per qu

A long time ago people learned something that would **alter** history: people learned how to **harness** fire. By striking stones together, a person could make a **spark**. Most likely, two **minerals** were used as **equipment** for starting fires. They gave off sparks when hit with something hard. The other **method** of creating fire was rubbing wooden sticks together. Just as your hands get warm when you rub them together, the **friction** of wood being rubbed together **generates** heat. **Tinder** would be put nearby to catch fire.

When a fire is lit, it creates light. The flame's color can tell you how hot the flame is and how much energy is being **released**. A bright blue flame is very hot and a dull yellow flame is cooler.

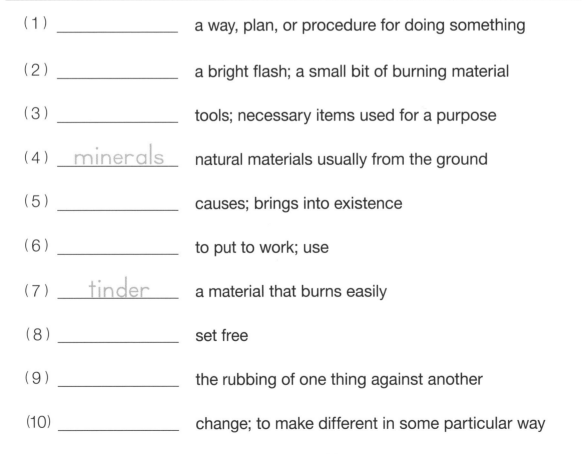

(1) _____ a way, plan, or procedure for doing something

(2) _____ a bright flash; a small bit of burning material

(3) _____ tools; necessary items used for a purpose

(4) _minerals_ natural materials usually from the ground

(5) _____ causes; brings into existence

(6) _____ to put to work; use

(7) _tinder_ a material that burns easily

(8) _____ set free

(9) _____ the rubbing of one thing against another

(10) _____ change; to make different in some particular way

Division

Date / /

Name

Level ★★

Score

/100

Math
DAY
9

1 Divide.

5 points per question

(1)

$$21\overline{)46} \quad 2\,R\square$$
$$\underline{4\,2}$$
$$\boxed{4}$$

(2)

$$21\overline{)67} \quad \square\,R\square$$

(3)

$$21\overline{)105} \quad \square$$

(4)

$$31\overline{)221}$$

(5)

$$32\overline{)235}$$

(6)

$$43\overline{)120}$$

(7)

$$47\overline{)300}$$

(8)

$$56\overline{)290}$$

(9)

$$53\overline{)350}$$

(10)

$$64\overline{)470}$$

(11)

$$67\overline{)600}$$

(12)

$$71\overline{)362}$$

(13)

$$76\overline{)532}$$

(14)

$$82\overline{)422}$$

(15)

$$88\overline{)528}$$

(16)

$$91\overline{)275}$$

(17)

$$94\overline{)658}$$

(18)

$$74\overline{)296}$$

(19)

$$85\overline{)455}$$

(20)

$$96\overline{)864}$$

Reading
DAY
9

Who, What, When, Where, Why & How
The Lion and the Mouse

Level ★
Score

Date
/ /

Name

1 Read the passage. Then choose words from the passage to answer the questions below.

100 poin for c

A mouse happened to run into the mouth of a sleeping lion, who awoke with a jolt. He pulled the frightened mouse from his mouth and was just about put him back in, when the little fellow began begging the lion to let him go. The mouse said, "If you spare my life, I shall be grateful forever and pay you back some day." The lion replied, "Haha! What good could a tiny mouse do—except to whet my appetite." But the lion thought the idea was so funny that he let the mouse go for giving him a good laugh and because the mouse was nothing more than a pre-snack snack to him.

Later that same day, the lion was running through the plains when he was caught by some hunters and bound by ropes to a tree. The mouse, hearing his roars and groans, came quickly. By gnawing the ropes, he was able to set the lion free, saying, "You laughed at me once, as if you could receive no return from me, but now, you see, it is you who have to be grateful to me." When there is a turn of events, even the most powerful can owe something to the weak.

(1) Who catches the lion?

Some _____ catch the lion.

(2) Where is the lion running when he is caught?

The lion is running through the _____.

(3) What is the lion doing when the mouse runs into his mouth?

The lion is _____ when the mouse runs into his mouth.

(4) When is the lion caught?

The lion is caught _____ that same day.

(5) Why does the mouse help the lion?

The mouse helps the lion because the lion spared the mouse's _____.

(6) How does the mouse help the lion escape?

The mouse helps the lion escape by _____ the ropes.

Division

Date   /   /

Name

Score   /100

1 Divide.

(1)
```
      1□R□
21)275
    21
     65
     □□
      □
```

(6)
```
36)989
```

(11)
```
      6□R□
21)1357
```

(16)
```
83)2713
```

(2)
```
      □□R□
21)665
```

(7)
```
43)500
```

(12)
```
41)2222
```

(17)
```
97)7324
```

(3)
```
27)631
```

(8)
```
44)903
```

(13)
```
53)3571
```

(18)
```
      □□□R□
23)4996
```

(4)
```
31)335
```

(9)
```
47)982
```

(14)
```
65)4456
```

(19)
```
33)7133
```

(5)
```
31)639
```

(10)
```
58)815
```

(15)
```
71)5248
```

(20)
```
42)8534
```

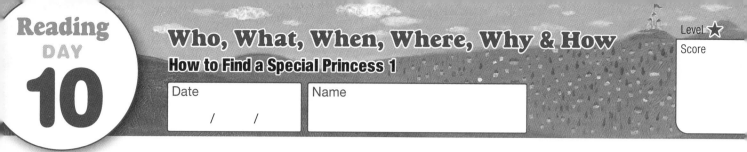

Who, What, When, Where, Why & How
How to Find a Special Princess 1

Level ⭐

Score

Date

/   /

Name

1 Read the passage. Then choose words from the passage to answer the questions below.

100 pc fo

> Once upon a time there was a prince who wanted to marry a princess, but she had to be a special princess. So he traveled east, then west, then north, and then south. There were plenty of princesses, but he could not find one that he considered special. In each case there was some little defect, which made him unsure. So he came home again in very low spirits. He thought he would be alone forever.
>
> The night after the prince's return there was a dreadful storm; there was thunder and lightning and the rain streamed down in torrents. It was fearsome! There was a knocking heard at the palace gate, and the old king and queen went to open it.
>
> There stood a princess outside the gate; but oh, she looked dreadful from the rain and the storm! The water was running down from her hair and her dress into the points of her shoes and out at the heels again. Her hair was a mess, whipping this way and that way from the wind. But she said she was a princess and had come to marry the prince.

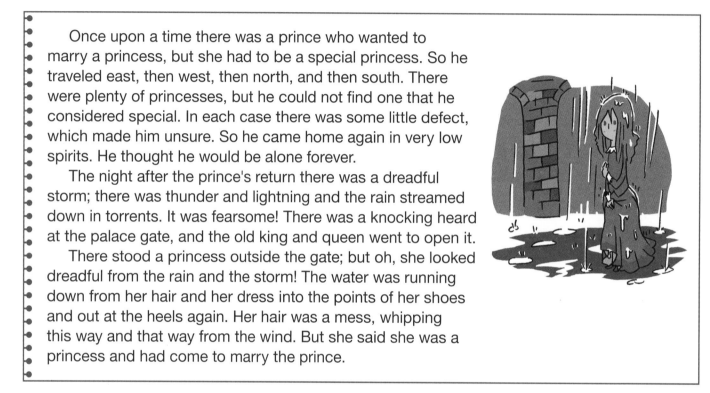

(1) Who answered the door of the palace?

   The _____ answered the door of the palace.

(2) What was the prince searching for?

   The prince was searching for a _____.

(3) Where did the prince travel?

   The prince traveled _____, then _____, then _____, and then _____.

(4) When was the storm?

   The storm was the night _____ the prince returned.

(5) Why was the prince in low spirits?

   The prince was in low spirits because he thought he would be _____forever.

(6) How did the princess look outside the gate?

   The princess looked _____.

Division

Level ★★★

Score

/100

Math
DAY
11

Date   /   /

Name

1 Divide.

10 points per question

(1)

$$121\overline{)537} \quad \square R\square\square$$

(5)

$$324\overline{)826}$$

(9)

$$832\overline{)9872}$$

(2)

$$173\overline{)888}$$

(6)

$$257\overline{)8784}$$

(10)

$$903\overline{)9879}$$

(3)

$$247\overline{)935}$$

(7)

$$391\overline{)8901}$$

(4)

$$307\overline{)935}$$

(8)

$$754\overline{)9004}$$

You're a math star!

Kumon Publishing Co.,Ltd.

21

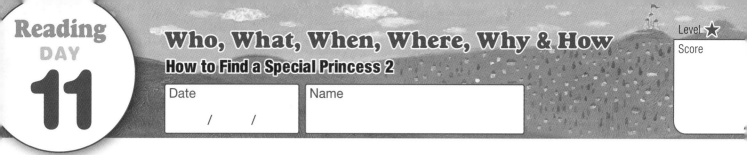

Reading DAY

11

Level ⭐
Score

Who, What, When, Where, Why & How
How to Find a Special Princess 2

Date
/   /

Name

① Read the passage. Then choose words from the passage to answer the questions below.   100 poi for

"Well, we shall soon find out!" thought the queen. But she said nothing and snuck into the guest bedroom. She took off all the bed linens and laid a pea on the mattress. Then she put twenty more mattresses on top of the pea and twenty quilts on the top of the mattresses. And this was the bed on which the princess was to sleep.

The next morning the queen asked how the princess had slept.

"Oh, very badly!" said the princess. "I scarcely closed my eyes all night! I don't mean to be ungrateful, but I don't know what was in the bed. I laid on something so hard that my whole body is black and blue. It is worse than the storm I escaped!"

Now they saw that she was a truly special princess because she had felt the pea through the twenty mattresses and the twenty quilts. Only a true princess could be so sensitive.

So the prince asked to marry her that very moment, and the pea was put into the Royal Museum, where it still can be seen today.

（1） Who laid the pea on the bed?

The _____ laid the pea on the bed.

（2） What was on top of the pea?

Twenty more _____ and twenty _____ were on top of the pea.

（3） Where did the princess sleep?

The princess slept in the _____ bedroom.

（4） When did the prince ask the princess to marry him?

The prince asked the princess to marry him _____.

（5） Why did the queen put the pea on the bed?

The queen put a pea on the bed to find out if the girl was a true _____.

（6） How did the prince know that she was a truly special princess?

The prince knew because only a true princess could be so _____.

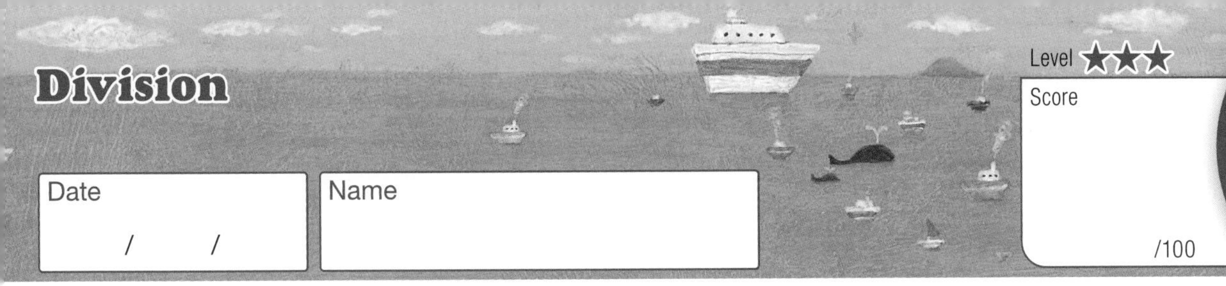

Division

Date / /

Name

Level ★★★

Score /100

Math DAY 12

1 Divide. 5 points per question

(1) 55)375	(6) 5)805	(11) 29)1506	(16) 328)1810
(2) 7)60	(7) 18)550	(12) 123)222	(17) 97)732
(3) 8)5846	(8) 63)9036	(13) 251)628	(18) 257)8993
(4) 38)621	(9) 9)7001	(14) 165)5348	(19) 482)7008
(5) 52)8061	(10) 42)720	(15) 18)5248	(20) 142)9500

Who, What, When, Where, Why & How
Discovering Vitamin K

Date
/ /

Name

① Read the passage. Then choose words from the passage to answer the questions below. 100 po for

---

Vitamins are needed by animals and plants for nutrition, growth, and life. Different vitamins have different jobs. Henrik Dam and Edward Doisy discovered vitamin K in 1934. Vitamin K is a vitamin that helps blood thicken and set. When blood sets, it forms a clot or lump and stops any bleeding.

Dam and his team discovered vitamin K by studying chicks that weren't well fed and bled easily. If the chicks had a cut it would also take a long time for the bleeding to stop. Dam believed that the chicks were missing a vitamin in their food that helped their blood clot. He found out this vitamin comes from green leaves and named it vitamin K. Dam and Doisy were able to find the vitamin in an alfalfa plant which has green leaves in a clover shape and blooms a blueish flower. They could make the chicks' blood clot better and faster by feeding them vitamin K.

In 1943, both scientists were awarded the Nobel Prize for Medicine for their research. This famous award is given each year in Stockholm, Sweden.

---

(1) Who discovered vitamin K?

_____ and _____ discovered vitamin K.

(2) What does vitamin K do?

Vitamin K helps blood _____ and _____.

(3) Where is the Nobel Prize given each year?

The Nobel Prize is given in _____, _____.

(4) When did Dam and Doisy discover vitamin K?

Dam and Doisy discovered vitamin K in _____.

(5) Why did the chicks bleed easily?

The chicks bled easily because they were missing _____.

(6) How could the scientists make the chicks' blood clot better and faster?

The scientists could make the chicks blood clot by _____ them vitamin K.

Decimals

Level ★★

Date / /

Name

Score /100

Math DAY 13

0 0.1 0.2 0.3 0.4 0.5 0.6 0.7 0.8 0.9 1 1.1 1.2

1 Write the correct decimal in each box. 4 points per box

(1)

0 0.1 [ ] 0.3 [ ] 0.5 [ ] 0.7 [ ] 0.9 1 1.1 1.2

(2)

0 [ ] 0.2 [ ] 0.4 [ ] 0.6 [ ] 0.8 [ ] 1 1.1 1.2

(3)

0 0.1 [ ] 0.3 0.4 0.5 [ ] 0.7 0.8 [ ] 1 [ ] 1.2

(4)

0 [ ] 0.2 0.3 0.4 [ ] 0.6 0.7 [ ] 0.9 [ ] 1.1 1.2

(5)

0 [ ] 1 [ ] [ ] 2 [ ]

(6)

0 [ ] 1 [ ] [ ] 2 [ ]

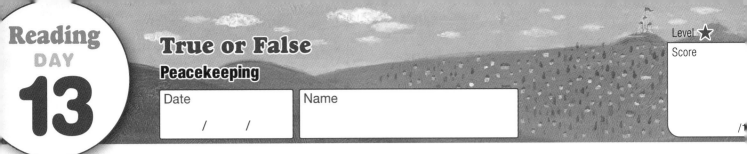

True or False
Peacekeeping

Date   /   /

Name

---

① Read the passage. Then read the sentences below. Circle the "T" if the sentence is true. Circle the "F" if the sentence is false.   10 points per que

> After World War II, many nations decided to form a group to assure people's freedom and to work towards peace around the world. On October 24, 1945, this group was formed and it was called the United Nations. Now that day is celebrated around the world as United Nations Day. The United Nations had many goals, but they started with four main aims: to set up and maintain international peace, to grow friendships between countries, to help countries work together to fix problems, and to convince nations to respect human rights and freedoms. Fifty-one countries joined together initially to create the United Nations. As of 2011, 193 nations are members.
>
> The United Nations has its own peacekeeping force, which includes members of the military, police, and general public who work to build peace in countries that have conflicts. Peacekeepers have been sent to countries all over the world to protect citizens and restore peace.

(1) Before World War II, many nations formed the United Nations.   T   F

(2) The United Nations was formed on October 24, 1945.   T   F

(3) October 24th is United Nations Day.   T   F

(4) The United Nations started with eight main goals.   T   F

(5) At first, fifty-one countries made up the United Nations.   T   F

(6) 190 nations were members as of 2011.   T   F

(7) One of the United Nations' goals is to grow friendships between countries.   T   F

(8) The peacekeepers are members of the military, police, judges and general public.   T   F

(9) The peacekeeping force builds peace in countries that have conflicts.   T   F

(10) Peacekeepers have been sent all over the world.   T   F

Decimals
Addition

Level ★★

Score

/100

Math
DAY
14

Date / /

Name

1 Add.

4 points per question

(1) $1 + 0.5 =$

(2) $2 + 0.7 =$

(3) $1 + 1.5 =$

(4) $1 + 0.9 =$

(5) $2 + 1.3 =$

(6) $0.2 + 0.7 =$

(7) $0.4 + 0.6 =$

(8) $0.6 + 1.7 =$

(9) $1.3 + 0.5 =$

(10) $1.5 + 0.7 =$

(11) $1.3 + 2 =$

(12) $1.9 + 1 =$

(13) $1.4 + 1.2 =$

(14) $2.1 + 1.5 =$

(15) $1.8 + 1.7 =$

2 Add.

4 points per question

(1)
$$\begin{array}{r} 1.3 \\ +\ 2.5 \\ \hline \end{array}$$

(2)
$$\begin{array}{r} 2.3 \\ +\ 0.4 \\ \hline \end{array}$$

(3)
$$\begin{array}{r} 14.5 \\ +\ 1.5 \\ \hline \end{array}$$

(4)
$$\begin{array}{r} 3.6 \\ +12.1 \\ \hline \end{array}$$

(5)
$$\begin{array}{r} 0.8 \\ +11.7 \\ \hline \end{array}$$

(6)
$$\begin{array}{r} 8.0 \\ +12.5 \\ \hline \end{array}$$

(7)
$$\begin{array}{r} 2.05 \\ +\ 1.60 \\ \hline \end{array}$$

(8)
$$\begin{array}{r} 0.68 \\ +\ 2.40 \\ \hline \end{array}$$

(9)
$$\begin{array}{r} 6.24 \\ +\ 1.75 \\ \hline \end{array}$$

(10)
$$\begin{array}{r} 5.52 \\ +14.48 \\ \hline \end{array}$$

Wow! You are mastering decimals!

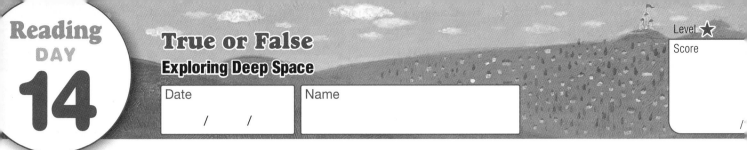

Reading
DAY
14

True or False
Exploring Deep Space

Date / /

Name

Level ⭐
Score

/

① Read the passage. Then read the sentences below. Circle the "T" if the sentence is true. Circle the "F" if the sentence is false. 10 points per que

> Remote-control spacecrafts have been flying around space for more than forty years. These explorers are "unmanned," meaning there is no person onboard the spacecraft. These crafts travel around like a stone from a slingshot. These amazing robots have gone as far as Mercury, Venus, Mars, Jupiter, Saturn, Uranus, and Neptune to get data and pictures. They are our eyes and ears in places where people cannot go. The spacecraft uses each planet's gravity to pull them in and shoot them onward. Gravity is the force that holds objects down on the surface of the earth.
> Yet none of these journeys would be possible without the Deep Space Network, which is a system of antennas. Antennas are devices that send and receive signals. These signals can travel up to billions of miles or kilometers. The farther a spacecraft has to go, the larger the antenna needs to be. Some antennas can be as large as a two-story house or even larger!

(1) Remote-control spacecrafts have been flying around space for more than four decades.　　T　　F

(2) Spacecrafts use a planet's gravity like a slingshot to pull them in and shoot them onward.　　T　　F

(3) Remote-control spacecrafts have gone as far as Neptune.　　T　　F

(4) Without the Deep Space Network, we would not be able to have so many television channels.　　T　　F

(5) Unmanned spacecrafts take pictures and get data.　　T　　F

(6) The Deep Space Network is a system of antennas.　　T　　F

(7) Remote-control spacecrafts always have at least one pilot on board.　　T　　F

(8) Antennas are devices that only receive signals.　　T　　F

(9) The farther a spacecraft has to go, the larger the antenna needs to be.　　T　　F

(10) Signals can travel only one million miles.　　T　　F

Decimals
Subtraction

Date   /   /

Name

1 Subtract.   4 points per question

(1) $0.8 - 0.3 =$ (6) $2.5 - 0.9 =$ (11) $2.1 - 0.3 =$

(2) $0.6 - 0.2 =$ (7) $1.7 - 0.7 =$ (12) $2.8 - 0.9 =$

(3) $1.5 - 0.2 =$ (8) $2.8 - 2 =$ (13) $3.5 - 1.7 =$

(4) $1.9 - 0.7 =$ (9) $3.5 - 1.3 =$ (14) $2.3 - 1.5 =$

(5) $1.4 - 0.6 =$ (10) $1.8 - 1 =$ (15) $3.2 - 2.8 =$

2 Subtract.   4 points per question

(1)
$$\begin{array}{r} 2.3 \\ - 1.1 \\ \hline \end{array}$$

(4)
$$\begin{array}{r} 2.6 \\ - 1.8 \\ \hline \end{array}$$

(7)
$$\begin{array}{r} 12.4 \\ - 3.4 \\ \hline \end{array}$$

(10)
$$\begin{array}{r} 4.54 \\ - 1.74 \\ \hline \end{array}$$

(2)
$$\begin{array}{r} 3.8 \\ - 2.5 \\ \hline \end{array}$$

(5)
$$\begin{array}{r} 3.3 \\ - 1.5 \\ \hline \end{array}$$

(8)
$$\begin{array}{r} 15.3 \\ - 7.0 \\ \hline \end{array}$$

Don't forget!
When you subtract decimals, line up the decimal points.

(3)
$$\begin{array}{r} 14.7 \\ - 0.4 \\ \hline \end{array}$$

(6)
$$\begin{array}{r} 4.2 \\ - 0.6 \\ \hline \end{array}$$

(9)
$$\begin{array}{r} 5.86 \\ - 3.50 \\ \hline \end{array}$$

Cause & Effect
The Peasant and the Bear 1

Date   /   /

Name

① Read the passage. Then answer the questions below.

> Once upon a time there was a peasant whose wife and children left him, and so he was all alone with no one to help him in his home or his fields. So he went to the bear and said, "Look here, Bear, let's plant our garden together."
>
> And the bear asked, "But how shall we divide it afterwards?"
>
> "How shall we divide it?" asked the peasant. "Well, you take all the tops and let me have all the roots."
>
> "All right, we have a deal," answered the bear.
>
> So they sowed some potatoes, and they grew beautifully. The bear worked hard and gathered all the potatoes. Then they began to divide them. The peasant said, "The tops are yours, aren't they, Bear?"
>
> "Yes," he answered.
>
> So the peasant cut off all the potato tops, which were only bitter leaves and gave them to the bear. Then the farmer sat down to count the delicious potatoes. The bear realized that the peasant outwitted him and he huffily went to his cave.

(1) Why was the peasant all alone?   30 points for co

The peasant was all alone because his _____ and _____ left him.

(2) Number the statements below in the order in which they occurred.   30 points for co

(   ) The bear asks how they will divide the food.

(   ) The peasant and the bear make a deal.

(   ) The peasant asks the bear to work together.

(   ) They grow potatoes and harvest them.

(3) Complete the chart with words from the passage above.   40 points for co

Cause	Effect
The peasant is alone.	He asks the bear to _____ a garden together.
They make a deal.	The peasant gets the _____, and the bear gets the _____.
They sow some _____.	The bear gets bitter _____ and the peasant gets delicious roots.

Improper Fractions

Date   /   /

Name

1 Rewrite the improper fractions as mixed numbers or whole numbers.

4 points per question

(1) $\dfrac{6}{5} = 1\dfrac{\Box}{5}$

(6) $\dfrac{5}{4} =$

(11) $\dfrac{16}{7} =$

(2) $\dfrac{8}{5} =$

(7) $\dfrac{9}{4} =$

(12) $\dfrac{13}{8} =$

(3) $\dfrac{13}{5} =$

(8) $\dfrac{7}{7} =$

(13) $\dfrac{19}{8} =$

(4) $\dfrac{6}{6} = \Box$

(9) $\dfrac{10}{7} =$

(14) $\dfrac{10}{9} =$

(5) $\dfrac{11}{6} =$

(10) $\dfrac{13}{7} =$

(15) $\dfrac{16}{9} =$

2 Rewrite the mixed numbers and whole numbers as improper fractions.

4 points per question

(1) $1 = \dfrac{\Box}{4}$

(5) $2\dfrac{1}{5} =$

(9) $1\dfrac{3}{7} =$

(2) $1\dfrac{2}{5} = \dfrac{\Box}{5}$

(6) $1\dfrac{1}{3} =$

(10) $1\dfrac{3}{9} =$

(3) $1\dfrac{1}{4} =$

(7) $2\dfrac{2}{3} =$

(4) $2 = \dfrac{\Box}{5}$

(8) $2 = \dfrac{\Box}{6}$

Reading
DAY
16

Cause & Effect
The Peasant and the Bear 2

Date   /   /

Name

Level ⭐

Score

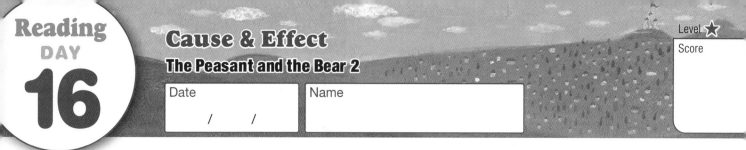

1 Read the passage. Then answer the questions below.

The next spring the peasant again came to see the bear and said, "Look here, Bear, let's work together again, shall we?"

The bear remembered the potato disaster from the year before and answered, "Right-ho! Only this time, I'll make the deal: you can have the tops, and I'm going to have the roots!"

"Very well," said the peasant.

But this year they sowed some wheat, and when the ears grew up and ripened, you never saw such a sight. The bear worked hard and gathered all the wheat, and then they began to divide it. This time, the peasant took all the tops with the grain for baking bread and gave the bear the straw and the roots, which weren't much good for anything. The bear realized that the peasant had outwitted him again!

"Well, good-bye!" said the bear to the peasant, "I'm not going to work with you anymore. You're too crafty!" And with that he went off into the forest.

(1) Why did the bear make a different deal?   20 poi

The bear made a different deal because he remembered the _____
from the year before.

(2) Number the statements below in the order in which they occurred.   40 poi
                                                                                              for
(   ) The bear gets the straw and roots.

(   ) They grow wheat together.

(   ) The bear makes a new deal.

(   ) The peasant asks the bear to work together again.

(3) Complete the chart with words from the passage above.   10 poi
                                                                                              per

Cause	Effect
The bear remembers last year's potato disaster.	He makes a new _____.
They make a deal.	The peasant gets the _____, and the bear gets the _____.
They plant wheat.	The bear gets the _____ and roots, and the peasant gets the grain.

Fractions
Addition

Date   /   /

Name

Level ★★

Score

/100

Math
DAY
17

1 Add.

5 $\frac{points}{per\ question}$

(1) $\dfrac{2}{5} + \dfrac{1}{5} = \dfrac{\boxed{}}{5}$

(2) $\dfrac{2}{5} + \dfrac{2}{5} =$

(3) $\dfrac{3}{5} + \dfrac{1}{5} =$

(4) $\dfrac{2}{7} + \dfrac{1}{7} =$

(5) $\dfrac{3}{7} + \dfrac{1}{7} =$

(6) $\dfrac{1}{7} + \dfrac{4}{7} =$

(7) $\dfrac{2}{7} + \dfrac{3}{7} =$

(8) $\dfrac{2}{9} + \dfrac{2}{9} =$

(9) $\dfrac{1}{9} + \dfrac{4}{9} =$

(10) $\dfrac{4}{9} + \dfrac{3}{9} =$

(11) $\dfrac{1}{5} + \dfrac{4}{5} = \dfrac{\boxed{}}{5} = \boxed{}$

(12) $\dfrac{2}{7} + \dfrac{4}{7} =$

(13) $\dfrac{3}{7} + \dfrac{4}{7} =$

(14) $\dfrac{1}{9} + \dfrac{7}{9} =$

(15) $\dfrac{4}{9} + \dfrac{5}{9} =$

(16) $\dfrac{2}{11} + \dfrac{6}{11} =$

(17) $\dfrac{3}{11} + \dfrac{5}{11} =$

(18) $\dfrac{4}{11} + \dfrac{7}{11} =$

(19) $\dfrac{5}{11} + \dfrac{4}{11} =$

(20) $\dfrac{2}{7} + \dfrac{5}{7} =$

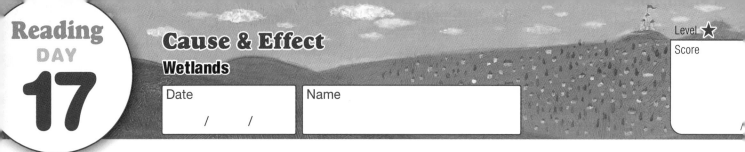

Cause & Effect
Wetlands

Date   /   /

Name

Level ★

Score

① Read the passage. Then answer the questions below.

"Wetland" is the name for any area of land that is between dry land and water, like swamps and bogs. When you think of wetlands you may think of mud, annoying mosquitoes, and stinky odors. People have destroyed many wetlands because they didn't know their value. More than half of the wetlands in the United States have been drained, filled, or used for the disposal of garbage. However, wetlands are a very important natural resource. Wetlands are similar to rain forests and coral reefs because they are home to many different animals, plants, and fish. Certain animals, like the wood stork, are endangered because of wetland destruction. Without wetlands, these animals will no longer be able to live. Wetlands also act like a sponge and soak up flooding water, rain, or melting snow, thereby protecting people and land.

Because wetlands are in danger, animals, plants, land, and people are in danger, too. Therefore, environmental groups are starting programs to save the wetlands. Some governments are also passing laws to protect these areas.

(1) Why have people destroyed many wetland areas?   20 points

   People have destroyed many wetland areas because people didn't know their

   _____.

(2) How are wetlands similar to rain forests and coral reefs?   20 points for com

   Wetlands are similar to rain forests and coral reefs because they are

   _____ **to many different** _____, _____, **and** _____.

(3) Why is the wood stork endangered?   20 points for con

   The wood stork is endangered because of _____ _____.

(4) Complete the chart with words from the passage above.   10 point per w

Cause	Effect
Wetlands soak up water.	Wetlands protect _____ and _____ from flooding.
The destruction of wetlands	Certain animals will no longer be able to _____.
Because wetlands are in _____.	Environmental groups are starting programs to save the wetlands.

Fractions
Subtraction

Date / /

Name

Level ★★

Score

/100

Math
DAY
18

1 Subtract.

5 points per question

(1) $\dfrac{2}{5} - \dfrac{1}{5} = \dfrac{\boxed{}}{5}$

(2) $\dfrac{4}{5} - \dfrac{2}{5} =$

(3) $\dfrac{4}{5} - \dfrac{1}{5} =$

(4) $\dfrac{2}{7} - \dfrac{1}{7} =$

(5) $\dfrac{3}{7} - \dfrac{1}{7} =$

(6) $\dfrac{4}{7} - \dfrac{2}{7} =$

(7) $\dfrac{6}{7} - \dfrac{2}{7} =$

(8) $\dfrac{4}{9} - \dfrac{2}{9} =$

(9) $\dfrac{5}{9} - \dfrac{1}{9} =$

(10) $\dfrac{7}{9} - \dfrac{5}{9} =$

(11) $1 - \dfrac{3}{5} = \dfrac{\boxed{}}{5}$

(12) $\dfrac{7}{5} - \dfrac{4}{5} =$

(13) $\dfrac{8}{7} - \dfrac{5}{7} =$

(14) $1 - \dfrac{3}{7} =$

(15) $\dfrac{8}{9} - \dfrac{5}{9} =$

(16) $\dfrac{10}{11} - \dfrac{6}{11} =$

(17) $1 - \dfrac{5}{8} =$

(18) $\dfrac{13}{11} - \dfrac{7}{11} =$

(19) $1 - \dfrac{4}{9} =$

(20) $1 - \dfrac{2}{11} =$

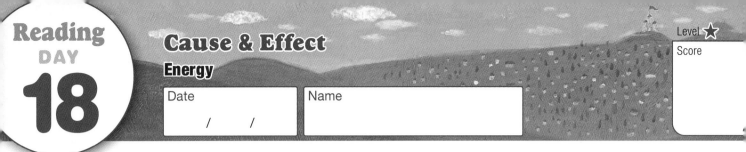

Cause & Effect
Energy

Date   /   /

Name

① Read the passage. Then answer the questions below.   25 points per qu

Most of our electricity comes from burning coal, oil, or gas, but there is a limited amount of these fuels, and one day there will be none left. Therefore, many people are looking for new ways to create electricity from resources that won't run out—in other words, a source of renewable energy.

People also want to find new types of energy because burning coal, oil, and gas pollutes the earth. Scientists are studying ways to make energy out of sunlight, wind power, and water.

The most common renewable source of electricity is hydropower. Hydropower is popular because it is not very expensive to produce. Hydropower can be created by using the water from a waterfall. Another way is by using a river and a special dam (a barrier that controls the flow of water in a river). When water from a river passes through a dam, the water turns a machine that looks like a fan. This movement creates energy that can be captured and turned into electricity. By controlling the flow of water, people can produce more or less electricity.

(1) Why is electricity made from burning coal not considered renewable energy?

Electricity made from burning coal is not considered renewable because there is

a _____ amount of coal.

(2) Why are people looking for new types of energy?

People are looking for new types of energy because coal, oil, and gas _____ the earth.

(3) Why is hydropower popular?

Hydropower is popular because it is not very _____ to produce.

(4) Complete the chart with words from the passage above.

Cause Effect	Effect
Limited amounts of coal, oil, and gas	People are looking for new ways to create _____ .
Water from a river goes through a hydropower dam	The water _____ a machine that looks like a fan
Controlling the flow of water	People can _____ more or less electricity.

36

Word Problems
Multiplication

Level ★★
Score

Math
DAY
19

/100

Date   /   /

Name

1 Read the word problem and write the number sentence below. Then answer the question. 20 points per question

(1) A box of pencils includes 12 pencils. If John bought 7 boxes, how many pencils did he buy?

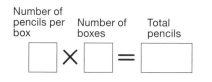

Number of pencils per box   Number of boxes   Total pencils

□ × □ = □

Ans. _____

(2) Robin's school has 25 classes. There are 30 students in each class. How many students are in Robin's school?

Ans. _____

(3) Colored pencils are sold in packs of 8. Your class has 28 packs. How many colored pencils does your class have?

Ans. _____

(4) The gardener gave each student 15 seeds to plant. If there are 27 students, how many seeds did the gardener give away?

Ans. _____

(5) Stickers are sold in rolls of 24. If Julie buys 11 rolls, how many stickers does she have?

Ans. _____

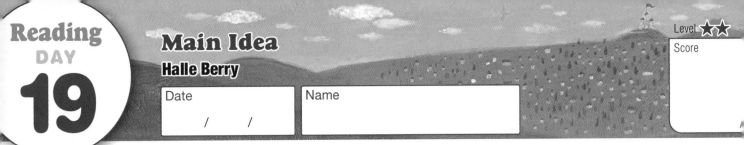

Main Idea
Halle Berry

Level ⭐⭐
Score

Date   /   /

Name

① Read the passage. Then answer the questions below.

Halle Berry is an admired public figure in American history. She was the first African-American woman to win an Academy Award for Best Actress. The Academy Awards are presented each year by the Academy of Motion Picture Arts and Sciences to recognize achievements in film.

Halle Berry was born on August 14, 1966, in Cleveland, Ohio. She was a teenage finalist in national beauty pageants and went on to work as a model. She began acting when she was twenty-three years old. She was cast in many different roles and earned a lot of praise for her work.

In 1999, Berry even starred in a film about the historic movie star Dorothy Dandridge, who was the first African American to be nominated for the Academy Award for Best Actress. When Berry won the Academy Award in 2001, she said, "This moment is so much bigger than me. This moment is for Dorothy Dandridge, Lena Horne, [and] Diahann Carroll," and she went on to dedicate the award to actresses of color.

(1) Read each title below. For which paragraph would each make a good title? Draw a line to connect each to the appropriate paragraph.   60 poin for c

    (a) Halle Berry's Childhood

    (b) Halle Berry's Tribute to
        African-American Actresses

    (c) Introduction to Halle Berry
        and the Academy Awards

    (i) First paragraph

    (ii) Second paragraph

    (iii) Third paragraph

(2) Put a check (✓) next to the best title for the whole passage below.   40 poin for c

    (   ) How to Break Into Acting

    (   ) The Biography of Halle Berry

    (   ) The Academy Awards

Don't forget! The **main idea** is a statement that expresses the most important information in the passage.

Word Problems

Division

Date   /   /

Name

1 Read the word problem and write the number sentence below. Then answer 20 points per question the question.

(1) We had 350 inches of ribbon for our group. We divided the ribbon equally among 7 students. How much ribbon did each student get?

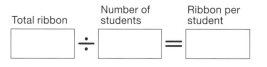

Total ribbon   Number of students   Ribbon per student

[   ] ÷ [   ] = [   ]

Ans. _____

(2) There are 204 roses. The florist split them evenly into 12 bunches. How many roses are in each bunch?

Ans. _____

(3) The cafeteria has 265 apples. They are divided into 8 boxes equally. How many apples are in each box, and how many apples remain?

Total apples   Number of boxes   Apples per box   Remaining

[   ] ÷ [   ] = [   ] R [   ]

Ans. _____

(4) Robin has 186 lollipops at his party. He divides them equally among 21 people. How many lollipops does each person get, and how many lollipops remain?

Ans. _____

(5) The art teacher has 113 sheets of colored paper. She divides them equally among 24 students. How many sheets of colored paper does each student get?

Ans. _____

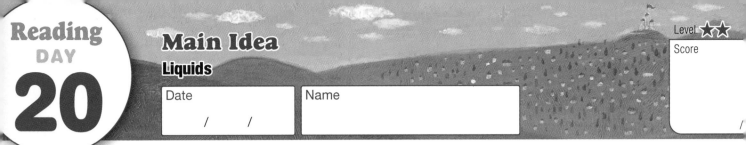

Main Idea
Liquids

Level ★★
Score

Date	Name
/ /	

1 Read the passage. Then answer the questions below.

A long time ago, Greeks believed that all liquids were made up of mostly water. However, scientists have discovered that all liquids are made up of particles called atoms. The smallest unit of water is a cluster of only three atoms.

Liquids can adapt to all different kinds of situations. Liquid can be thinly spread out, like when it spills across a table; or it can be tightly packed together, like when it is held in a bottle. When liquid is heated, the spaces between the particles expand and, so does the liquid. The opposite also occurs when a liquid is cooled—the particles contract and get closer together.

The tiny particles that make up a liquid are also attracted to each other and tend to keep close together. This attraction creates tension between the particles, which is why when you fill up a cup to the top, the water at the surface holds tight like the skin of a balloon. This tension also allows very light insects to walk on water. This phenomenon is called surface tension.

Liquids are also very powerful. Given enough time, liquids can wear away solid surfaces, like rocks. For example, a canyon is a deep and steep valley that has been carved out by a river. These kinds of valleys are often located where the river has a strong current that runs rapidly.

(1) Read each title below. For which paragraph would each make a good title? Draw a line to connect each title to the appropriate paragraph.   **40** points for cc

(a) The Discovery of Atoms   (i) First paragraph

(b) Surface Tension   (ii) Second paragraph

(c) The Power of Liquids   (iii) Third paragraph

(d) Liquid in Different Forms   (iv) Fourth paragraph

(2) Put a check (✓) next to the sentence that describes the main message of the fourth   **30** points
paragraph.
(   ) Liquids can adapt.   (   ) Liquids are powerful
(   ) Canyons are made by liquid.   (   ) When liquid cools, the particles contrast.

(3) What is the main idea of the whole passage? Put a check (✓) next to the correct idea   **30** points
below.
(   ) Liquids are made up of atoms.   (   ) Liquids are unchanging.
(   ) Liquids have adaptable qualities.   (   ) The Greeks didn't know about atoms.

Word Problems

Division

Level ★★

Score

/100

Math
DAY
21

Date   /   /

Name

1 Read the word problem and write the number sentence below. Then answer 20 points per question
the question.

(1) Mary has 208 flowers. If she puts 8 flowers into each vase, how many vases will she need?

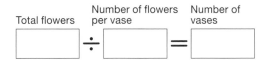

Total flowers | Number of flowers per vase | Number of vases

[   ] ÷ [   ] = [   ]

Ans. _____

(2) An art class has a piece of string that is 194 inches long. The teacher cuts it and makes 16-inch segments of string. How many segments will there be, and how long is the remaining string?

Ans. _____

(3) Mother bought 105 apples. She put 6 apples in each bag. How many bags are there, and how many apples remain?

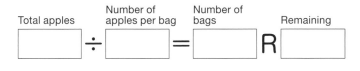

Total apples | Number of apples per bag | Number of bags | Remaining

[   ] ÷ [   ] = [   ] R [   ]

Ans. _____

(4) Tina has 376 inches of ribbon. She divides it into sections that are 14 inches long. How many sections of ribbon will she have, and how long is the remaining piece of ribbon?

Ans. _____

(5) Grandmother has 310 candies to send to her family. She puts 28 candies into each box. How many boxes will she need, and how many candies will be left over?

Ans. _____

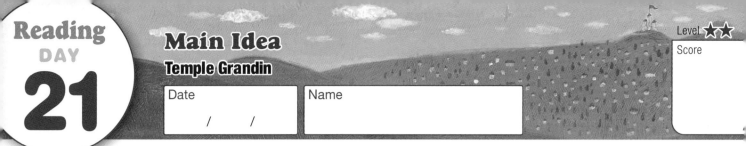

Main Idea
Temple Grandin

Date   /   /

Name

① Read the passage. Then answer the questions below.

Temple Grandin was born on August 29, 1947, in Boston, Massachusetts. Grandin wasn't able to talk until age three. Her doctors diagnosed her as autistic. A person with autism often finds it difficult to interact and communicate with other people.

Although Grandin faced many challenges because of autism, her parents nurtured her intelligence, and she eventually went on to speak, finish high school, and study psychology in college in New Hampshire. Afterwards, she earned a master's degree and a doctorate in animal science, which was very uncommon for a woman at that time.

Because of her disability, Grandin devoted her life to learning about anxiety in people and animals and finding solutions. Grandin experienced a lot of anxiety because autistic people can be very sensitive to sound and touch. While still in high school, she designed a "squeeze machine" to help relieve her nervousness. The machine was modeled after a chute that held animals in place. Grandin's invention is now used with autistic children and adults.

However, Grandin is most well known for her innovative work with animals. She has designed humane, or more gentle, livestock facilities that eliminate pain and fear in animals. Her designs also allow workers to move animals without frightening them. She has also written several books about animal behavior.

(1) Read each title below. For which paragraph would each make a good title? Draw a line to connect each title to the appropriate paragraph.

40 poin. for c

(a) Grandin's First Invention   (i) First paragraph

(b) Being Diagnosed with Autism   (ii) Second paragraph

(c) Grandin's Legacy with Animals   (iii) Third paragraph

(d) Grandin Succeeds at School   (iv) Fourth paragraph

(2) What is the main idea of the whole passage? Put a check (✓) next to the correct idea below.

30 poin.

(   ) People with autism have problems interacting with others.

(   ) Temple Grandin faced many challenges.

(   ) Temple Grandin was inspired by her own challenges to help anxious animals and people.

30 poin.

(3) What detail supports the main idea? Put a check (✓) next to the answer.

(   ) Temple Grandin wasn't able to talk until age three.

(   ) Temple Grandin was born on August 29, 1947.

(   ) Temple Grandin studied pyschology and animal science.

You're a great reader!

Word Problems
Division

Level ★★

Score

/100

Math
DAY
22

Date / /

Name

1 Read the word problem and write the number sentence below. Then answer the question. 20 points per question

(1) Kate has 54 dimes for her collection. Her younger sister has 18 dimes. How many times more dimes does Kate have than her sister?

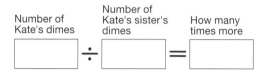

Number of Kate's dimes ÷ Number of Kate's sister's dimes = How many times more

Ans.

(2) The grocer has 112 oranges in a box and 14 oranges out front. How many times more oranges does he have in the box than he has out front?

Ans.

(3) A red ribbon is 36 yards long. It is 3 times longer than the blue ribbon. How long is the blue ribbon?

Length of red ribbon ÷ How many times longer = Length of blue ribbon

Ans.

(4) The bear at the zoo weighs 560 pounds. He weighs 7 times as much as the chimpanzee. How much does the chimpanzee weigh?

Ans.

(5) The bedroom in Smith's house is 192 square feet, and the closet is 32 square feet. How many times bigger is the bedroom than the closet?

Ans.

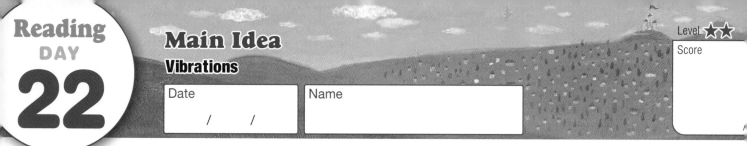

Reading
DAY
22

Main Idea
Vibrations

Date   /   /

Name

Level ★★
Score

① Read the passage. Then answer the questions below.

Did you know that glass can be shattered with only the force of the human voice? This is because of the power of vibrations, or movement back and forth. The number of vibrations that an object makes each second is called its natural frequency. Anything that can vibrate — everything from a bridge to a violin string—has its own natural frequency. Just like a swing in a playground, if an object is given a push, it will move back and forth at its natural frequency and then gradually stop. But if you continue to push a swing in the right rhythm, it can rise higher and higher. This happens when you push according to the swing's natural frequency.

The same thing happens with the shattering glass. When someone sings, his or her voice creates a sound wave that vibrates. Different notes make different sound waves and thus vibrate at different rates. If a note is sung with a rate of vibration that matches the natural frequency of the glass, the glass could shatter. When the vibrations match, the energy from the voice transfers to the glass, and the powerful vibrations destroy the glass. This transfer is called "resonance."

Luckily, there is a way to demonstrate resonance without destroying glasses. By taking a wine glass and running a wet finger quickly around the rim of the glass, a person can create a note. If the person sings the same note aloud, the glass will resonate the note and the sound will become slightly louder.

(1) Read each title below. For which paragraph would it make a good title? Draw a line to connect each title to the appropriate paragraph.   40 points per co

(a) How a Note Can Destroy a Glass   (i) First paragraph

(b) An Experiment to Show Resonance   (ii) Second paragraph

(c) Introduction to Vibration   (iii) Third paragraph

(2) What is the main idea of the whole passage? Put a check (✓) next to the correct idea below.   30 points

(   ) Science experiments are fun and informative.

(   ) Vibration and resonance can be a powerful force together.

(   ) Everything has its own natural rate of vibration.

(3) What detail supports the main idea? Put a check (✓) next to the answer.   30 points

(   ) If you continue to push a swing at its natural frequency, it will rise.

(   ) Many things—from bridges to violin strings—vibrate.

(   ) Singing loudly is not good for glasses.

Word Problems
Addition of Decimals

Date / /

Name

Level ★★

Score

/100

Math
DAY
23

1 Read the word problem and write the number sentence below. Then answer the question.   20 points per question

(1) You used 0.2 pounds of sugar in your cake, and 1.7 pounds of sugar are left over. How much sugar was there in the beginning?

Ans. _____

(2) Julian's bag weighs 2.8 pounds. His father's bag is 1.2 pounds heavier than his. How much does his father's bag weigh?

Ans. _____

(3) Dan and Wendy were trying to throw a big rock. Dan threw it 1.3 meters. Wendy threw it 70 centimeters farther. How far did Wendy throw the rock?

$70\,cm = 0.7\,m$

Ans. _____

(4) Ava's bag weighs 2.4 kilograms. Her sister's bag is 600 grams heavier. How much does her sister's bag weigh?

$600\,g = 0.6\,kg$

Ans. _____

(5) Kelly had 2.1 liters of water in her water bottle. Her big water bottle can hold 800 milliliters more. How much can her big water bottle hold?

$800\,mL = 0.8\,L$

Ans. _____

Reading
DAY
23

Main Idea
Roberto Clemente

Level ★★
Score

Date   /   /

Name

① Read the passage. Then answer the questions below.

> Roberto Clemente was one of the first Latin American baseball stars. He was born in a modest house in Puerto Rico on August 18, 1934. Clemente went on to become a twelve-time All Star and do important charity work in his free time.
>
> At only fourteen years old, Clemente began playing softball on a men's team. By eighteen, he turned professional. In February of 1954, the Brooklyn Dodgers recruited Clemente but placed him in the minor leagues, where he didn't play very often. The Dodgers tried to hide his talent so other teams wouldn't want him. But it was too late—the Pittsburgh Pirates brought Clemente up to the major leagues.
>
> Over eighteen seasons, Clemente collected impressive statistics and delighted baseball fans. No matter what kind of pitch, he could hit the ball. He had lightning speed, which made him a great base runner. He was also well known for his powerful and accurate throwing arm. Even towards the end of his career, Clemente continued to set records.
>
> But one of the biggest challenges Clemente faced was racial prejudice. Many baseball fans, reporters, and players were rude or nasty to Clemente because he was black and Latino. However, he always defended his rights and the rights of others. Clemente said, "My greatest satisfaction comes from helping to erase the old opinion about Latin Americans and blacks."
>
> On December 31, 1972, Clemente died in a plane crash only a few miles from where he was born. He was on his way to deliver aid to earthquake victims in Nicaragua. He was only thirty-eight, but he had become a baseball legend.

(1) Complete the main ideas for each paragraph in the chart below.   **70** poir for e

Paragraph	Main idea
First paragraph	Introduction to _____ _____.
Second paragraph	The start of Clemente's _____ career.
Third paragraph	Clemente ____ baseball records and delighted _____.
Fourth paragraph	Clemente faced racial _____.
Fifth paragraph	Clemente died young but had become a baseball _____.

(2) What is the main idea of the whole passage? Put a check (✓) next to the correct idea   **30** poir
below.

(   ) Over eighteen seasons, Clemente set impressive records.

(   ) Clemente helped many people and fought racial prejudice.

(   ) Clemente overcame many challenges, became a baseball legend and helped people.

Word Problems
Subtraction of Decimals

Date	Name	Level ★★ Score
/ /		/100

Math
DAY
24

1 Read the word problem and write the number sentence below. Then answer the question. 20 points per question

(1) Selena wrapped presents for her friends. She used 1.6 yards of ribbon out of the 2.3 yards of ribbon she had. How much ribbon was left?

Ans. _____

(2) My mother had 3 pounds of flour. She used 0.3 pound to bake a cake. How much flour is left?

Ans. _____

(3) We have 2.4 liters of orange juice in the fridge. We also have apple juice, but 500 milliliters less. How much apple juice do we have in the fridge?

500 mL = 0.5 L

Ans. _____

(4) Brad and Mark both have sticks. Brad's is 2.1 meters long. If Brad's stick is 50 centimeters longer than Mark's, how long is Mark's stick?

50 cm = 0.5 m

Ans. _____

(5) My bag of pears is 0.8 kilograms heavier than your bag of pineapples. If my bag of pears weighs 2 kilograms, how much does your bag of pineapples weigh?

Ans. _____

Characters
The Tale of Mr. Tod

Date   /   /

Name

① Read the excerpt from *The Tale of Mr. Tod* by Beatrix Potter. Then answer the   25 poir per
questions below using words from the passage.

I have made many books about well-behaved people. Now, for a change, I am going to make a story about two disagreeable people, called Tommy Brock and Mr. Tod. Nobody could call Mr. Tod "nice." The rabbits could not bear him; they could smell him half a mile off. He was of a wandering habit and he had foxey whiskers; they never knew where he would be next.

One day he was living in a stick-house in the coppice, causing terror to the family of old Mr. Benjamin Bouncer. Next day he moved into a pollard willow near the lake, frightening the wild ducks and the water rats.

In winter and early spring he might generally be found in an earth amongst the rocks at the top of Bull Banks, under Oatmeal Crag.

He had half a dozen houses, but he was seldom at home.

The houses were not always empty when Mr. Tod moved out; because sometimes Tommy Brock moved in; (without asking leave).

Tommy Brock was a short, bristly, fat, waddling person with a grin; he grinned all over his face. He was not nice in his habits. He ate wasp nests and frogs and worms; and he waddled about by moonlight, digging things up.

His clothes were very dirty; and as he slept in the daytime, he always went to bed in his boots.

（1）What type of characters will the author write a story about?

The author will write about two _____ people.

（2）Put a check (✓) next to the words that describe Mr. Tod.
(   ) nice   (   ) unmoving   (   ) smelly
(   ) wanderer   (   ) never home   (   ) musical
(   ) animal-friendly   (   ) unpredictable   (   ) funny

（3）Put a check (✓) next to the words that describe Tommy Brock.
(   ) smiley   (   ) tall   (   ) short
(   ) nice   (   ) dirty   (   ) talented
(   ) normal   (   ) strange   (   ) smart

（4）Despite it being daytime, what did Tommy Brock do?

Tommy Brock _____ in the daytime.

Phenomenal work!

Word Problems
Mixed Calculations

Date / /

Name

Level ★★
Score

/100

Math
DAY
25

1 Read the word problem and write the number sentence below. Then answer 20 points per question the question.

(1) At the supermarket, my mother bought some meat for $14 and some vegetables for $12. If she paid with a $50-bill, how much change did she get? Using parentheses, write this down in a formula and then solve it.

☐ − (☐ + ☐) = ☐

Ans. _____

(2) Rina's book about animals has 256 pages. She read 64 pages yesterday and 57 pages today. How many pages are left?

Ans. _____

(3) Jack saw a jacket that he wanted to buy. He bargained with the store owner, who discounted the price $5. The jacket was originally $78, and Jack paid with a $100 bill. How much change will Jack get?

☐ − (☐ − ☐) = ☐

Ans. _____

(4) Because a dress was not popular, the shopkeeper discounted it $12. It used to cost $68. Helen bought it and paid with a $100 bill. How much change did she get?

Ans. _____

(5) You went shopping and bought a sandwich for $7 and a book for $12. If you paid with a $20 bill, how much change did you get?

Ans. _____

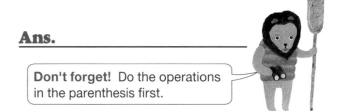

Don't forget! Do the operations in the parenthesis first.

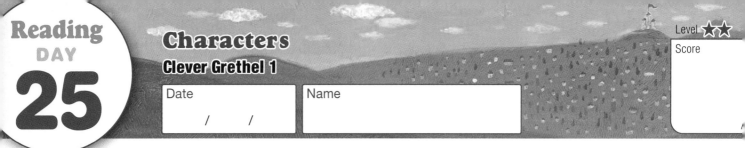

Reading
DAY
25

Characters
Clever Grethel 1

Level ★★
Score

Date   /   /

Name

1 Read the passage. Then answer the questions below.

> There was once a cook called Grethel, who wore shoes with red heels, and when she went out in them she gave herself great airs, and thought herself very fine indeed. When she came home again, she would take a drink of wine to refresh herself, and as that gave her an appetite, she would take some of the best of whatever she was cooking, until she had had enough;—"For," said she, "a cook must know how things taste."
>
> Now it happened that one day her master said to her,—
>
> "Grethel, I expect a guest this evening; you must make ready a pair of fowls."
>
> "Certainly, sir, I will," answered Grethel. So she killed the fowls, cleaned them, and plucked them, and put them on the spit, and then, as evening drew near, placed them before the fire to roast. And they began to be brown, and were nearly done, but the guest had not come.

(1) Put a check (✓) next to the words that describe Grethel.   60 points for cor

  (   ) angry   (   ) greedy   (   ) vain

  (   ) cook   (   ) modest   (   ) dirty

  (   ) generous   (   ) stem   (   ) curious

> And now they began to smell so good that Grethel saying, "I must find out whether they really are all right," licked her fingers, and then cried, "Well, I never! The fowls are good; it's a sin and a shame that no one is here to eat them!"
>
> So she ran to the window to see if her master and his guest were coming, but as she could see nobody she went back to her fowls. "Why, one of the wings is burning!" she cried presently, "I had better eat it and get it out of the way." So she cut it off and ate it up, and it tasted good, and then she thought, "I had better cut off the other too, in case the master should miss anything." And when both wings had been disposed of she went and looked for the master, but still he did not come. "Who knows," said she, "whether they are coming or not? They may have put up at an inn." And after a pause she said again, "Come, I may as well make myself happy, and first I will make sure of a good drink and then of a good meal."

(2) Why does Grethel say she must eat the first wing?   20 points

Grethel must eat the first wing because it is _____.

(3) Why does Grethel say she should eat the whole meal?   20 points

Grethel says she should eat the whole meal because she may as well make

herself _____.

Word Problems
Mixed Calculations

Level ★★
Score

/100

Math
DAY
26

Date	Name
/ /	

1 Read the word problem and write the number sentence below. Then answer the question.

20 points per question

(1) Alison wants to have 4 candies for everyone at her party. There are 11 boys and 13 girls at her party. How many candies will she need? Remember to use a formula.

Ans. _____

(2) Jack is buying food for his chicken and chicks. Each day, the chicks eat 30 seeds and the chicken eats 60 seeds. If Jack wants to feed them for a week, how many seeds must he buy?

Ans. _____

(3) Olive found a brush and paint set that she liked. The brush cost $5 and the paint cost only $3. How many sets can she buy if she has $96?

Ans. _____

(4) Gayle and her 2 brothers gathered their money and bought a $43 racing video game and a $44 basketball video game. What did each person pay?

Ans. _____

(5) In the pantry, the 2 cans of beans weigh 350 grams each, and the 4 cans of beets weigh 430 grams each. What is the total weight of the cans?

Ans. _____

Reading DAY 26

Characters

Clever Grethel 2

Level ★★

Score

Date / /

Name

① Read the passage. Then identify the statements as T (true) or F (false) according to the passage. 100 pc fo

Just as she was in the middle of it the master came back. "Make haste, Grethel," he cried, "the guest is coming directly!" "Very well, master," she answered, "it will soon be ready." The master went to see that the table was properly laid, and, taking the great carving knife with which he meant to carve the fowls, he sharpened it upon the step.

Presently came the guest, knocking very genteelly and softly at the front door. Grethel ran and looked to see who it was, and when she caught sight of the guest she put her finger on her lip saying, "Hush! Make the best haste you can out of this, for if my master catches you, it will be bad for you; he asked you to come to supper, but he really means to cut off your ears! Just listen how he is sharpening his knife!"

The guest, hearing the noise of the sharpening, made off as fast as he could go. And Grethel ran screaming to her master. "A pretty guest you have asked to the house!" she cried.

"How so, Grethel? What do you mean?" he asked.

"What indeed!" she said. "Why, he has gone and run away with my pair of fowls that I had just dished up."

"That's a pretty sort of conduct!" said the master, feeling very sorry about the fowls. "He might at least have left me one, that I might have had something to eat." And he called out to him to stop, but the guest made as if he did not hear him. Then he ran after him, the knife still in his hand, crying out, "Only one! Only one!" meaning that the guest should let him have one of the fowls and not take both. But the guest thought he meant to have only one of his ears, and he ran so much the faster that he might get home with both of them safe.

（1） The master sharpened his knife to cut the fowl.　　　T　　F

（2） Grethel told the guest the truth.　　　T　　F

（3） The guest believed Grethel.　　　T　　F

（4） Grethel was lying to the guest about her master.　　　T　　F

（5） Grethel told her master a lie.　　　T　　F

（6） Grethel found a clever way to cover up the eaten meal.　　　T　　F

Word Problems
Mixed Calculations

Date
/ /

Name

Level ★★
Score
/100

Math
DAY
27

1 Read the word problem and write the number sentence below. Then answer the question. 20 points per question

(1) Kate lost her dog. She made 250 flyers and gave 5 flyers each to 34 people. How many flyers did she have left over?

Ans. _____

(2) Andy saved up $37. Today his aunt gave him $50 and told him to divide it with his sister evenly. How much money does Andy have now?

Ans. _____

(3) A store has 1,000 gumballs. 820 gumballs weighs 2 kilograms. If you buy 1 kilogram of gumballs, how many gumballs will be left?

Ans. _____

(4) A store has 200 oranges divided equally in 5 crates. Tina bought 1 crate of oranges. Tina also bought 3 crates of apples. Each crate holds 30 apples. How many pieces of fruit did Tina buy?

Ans. _____

(5) Dana divided 72 stickers into 24 gift bags equally. Barry took 4 bags home. How many stickers did Barry take home?

Ans. _____

Actions & Descriptions
Black Beauty 1

Date / /

Name

Level ★★

Score

(1) Read the excerpt from *Black Beauty* by Anna Sewell. Then answer the questions below.

One night, a few days after James had left, I had eaten my hay and was lying down in my straw fast asleep, when I was suddenly roused by the stable bell ringing very loud. I heard the door of John's house open, and his feet running up to the hall. He was back again in no time; he unlocked the stable door, and came in, calling out, "Wake up, Beauty! You must go well now, if ever you did;" and almost before I could think he had got the saddle on my back and the bridle on my head. He just ran round for his coat, and then took me at a quick trot up to the hall door. The squire stood there, with a lamp in his hand.

"Now, John," he said, "ride for your life—that is, for your mistress' life; there is not a moment to lose. Give this note to Dr. White; give your horse a rest at the inn, and be back as soon as you can."

John said, "Yes, sir," and was on my back in a minute.

The gardener who lived at the lodge had heard the bell ring, and was ready with the gate open, and away we went through the park, and through the village, and down the hill till we came to the toll-gate. John called very loud and thumped upon the door; the man was soon out and flung open the gate.

"Now," said John, "do keep the gate open for the doctor; here's the money," and off he went again.

(1) What was Black Beauty doing when the stable bell rang?   15 points for co

Black Beauty was _____ down and _____ in his straw.

(2) What did John do to get Black Beauty ready to ride?   15 points for co

John _____ the _____ on his back and the _____ on his head.

(3) What did the squire tell John he must do?   15 points

The squire told John to _____ a note to Dr. White.

(4) Describe the route that John and Black Beauty took to get to the toll-gate.   15 points for co

John and Black Beauty went _____ the park, and the village, and _____ the hill.

(5) How did John call for the man at the toll-gate?   15 points

John called very _____ for the man.

(6) Put a check (✓) next to the words that describe the scene.   25 points for completion

( ) alarming   ( ) lazy   ( ) humorous

( ) relaxed   ( ) important   ( ) sad

( ) urgent   ( ) tense   ( ) boring

Don't forget! Actions are things that happen. **Descriptions** explain what something is or what something is like.

Word Problems
Mixed Calculations

Level ★★

Score

/100

Math
DAY
28

Date
/   /

Name

1 Read the word problem and write the number sentence below. Then answer the question. 20 points per question

(1) The teacher brought 21 dozen colored pencils to art class today. If she divided the pencils equally among 28 students, how many pencils did each student get?

Ans. _____

(2) Glen bought 6 packs of 45 stickers and 9 packs of 70 stickers. How many stickers did he get in all?

Ans. _____

(3) Eddie bought 5 big bags of chips for $15. His brother bought 3 small bags and paid $6. How much more expensive were Eddie's bags of chips?

Ans. _____

(4) My piggy bank has 37 coins in it. They are all pennies and nickles. If there are 5 more pennies than nickles, how many nickles do I have?

Ans. _____

(5) Shannon has 84 stamps and her sister has 50. How many stamps does Shannon have to give her sister so that they have the same amount?

Ans. _____

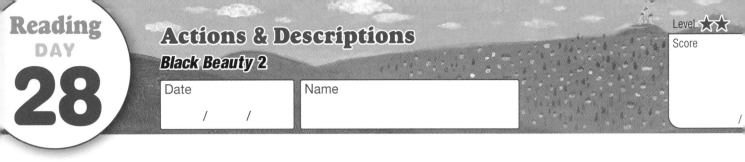

Actions & Descriptions
Black Beauty 2

Level ★★
Score

Date / /

Name

/

1 Read the passage. Then answer the questions below.

20 points per que

There was before us a long piece of level road by the river side; John said to me, "Now, Beauty, do your best," and so I did; I wanted no whip nor spur, and for two miles I galloped as fast as I could lay my feet to the ground; I don't believe that my old grandfather, who won the race at Newmarket, could have gone faster. When we came to the bridge John pulled me up a little and patted my neck. "Well done, Beauty! Good old fellow," he said.

He would have let me go slower, but my spirit was up, and I was off again as fast as before. The air was frosty, the moon was bright; it was very pleasant. We came through a village, then through a dark wood, then uphill, then downhill, till after eight miles' run we came to the town, through the streets and into the marketplace. It was all quite still except the clatter of my feet on the stones—everybody was asleep. The church clock struck three as we drew up at Dr. White's door.

(1) What did Black Beauty do for two miles?

Black Beauty _____ as fast as he could for two miles.

(2) When they arrived at the bridge, what did John do?

John _____ Black Beauty up and _____ his neck.

(3) Put a check (✓) next to the word that describes Black Beauty's run.

( ) relaxed   ( ) sluggish   ( ) fast

(4) Describe the route that John and Black Beauty took after the village.

John and Black Beauty went through a _____ wood, then _____, then

_____.

(5) Write a **D** next to the sentence below that is description only.

( ) There was a long piece of level road by the river side.

( ) John thumped upon the door.

( ) I galloped as fast as I could for two miles.

( ) The moon was bright.

( ) I was off again as fast as before.

( ) It was very pleasant.

( ) It was all quite still except the clatter of my feet on the stones.

( ) The air was frosty.

Word Problems
Mixed Calculations

Level ★★★

Score

/100

Math
DAY
29

Date

/ /

Name

1 Read the word problem and write the number sentence below. Then answer the question. 20 points per question

(1) Jessica's mother weighs 54 kilograms, and that is twice as much as Jessica weighs. Jessica weighs 3 times as much as her baby sister. How much does Jessica's baby sister weigh?

Ans. _____

(2) There are apples and melons in the fruit basket in the cafeteria. Altogether there are 35 pieces of fruit. If there are 4 times as many apples as melons, how many of each kind of fruit is in the basket?

Ans. Apples _____ Melons _____

(3) At Farmer William's farm, he has cows and horses. He has 3 times as many horses as cows. If there are 34 more horses than cows, how many of each does Farmer William have?

Ans. Cows _____ Horses _____

(4) Ted has 5 pieces of 8-inch tape, but he connects them so 2 inches overlap. How long is his new piece?

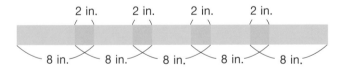

2 in.   2 in.   2 in.   2 in.

8 in.   8 in.   8 in.   8 in.   8 in.

Ans. _____

(5) Sally's living room is 3 meters and 50 centimeters wide. Her pictures are 35 centimeters wide, and she wants the pictures spaced equally as shown below. How much space should she put between the pictures in her living room?

35 cm

Ans. _____

Actions & Descriptions
Black Beauty 3

Date   /   /

Name

① Read the passage. Then answer the questions below.

25 points per q

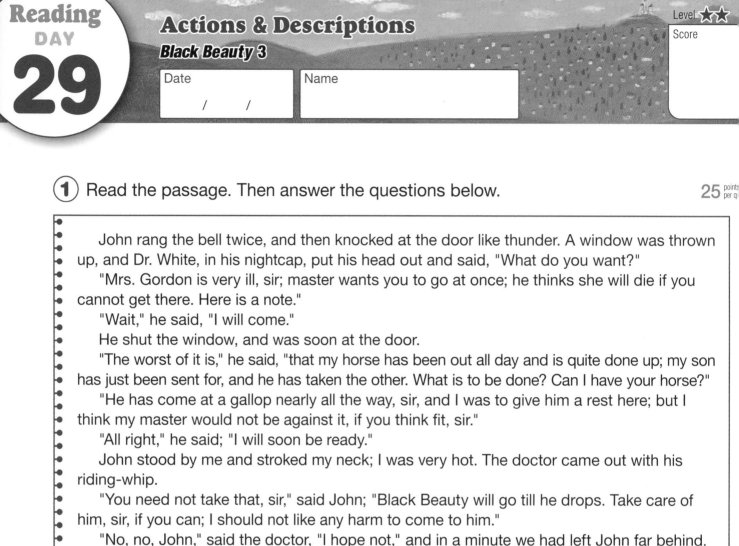

John rang the bell twice, and then knocked at the door like thunder. A window was thrown up, and Dr. White, in his nightcap, put his head out and said, "What do you want?"

"Mrs. Gordon is very ill, sir; master wants you to go at once; he thinks she will die if you cannot get there. Here is a note."

"Wait," he said, "I will come."

He shut the window, and was soon at the door.

"The worst of it is," he said, "that my horse has been out all day and is quite done up; my son has just been sent for, and he has taken the other. What is to be done? Can I have your horse?"

"He has come at a gallop nearly all the way, sir, and I was to give him a rest here; but I think my master would not be against it, if you think fit, sir."

"All right," he said; "I will soon be ready."

John stood by me and stroked my neck; I was very hot. The doctor came out with his riding-whip.

"You need not take that, sir," said John; "Black Beauty will go till he drops. Take care of him, sir, if you can; I should not like any harm to come to him."

"No, no, John," said the doctor, "I hope not," and in a minute we had left John far behind.

(1) What did John do to get the doctor to wake up?

John _____ the bell twice and _____ at the door.

(2) Describe how John knocked at the door.

John knocked at the door like _____.

(3) While they waited for the doctor, what did John do?

John _____ Black Beauty's neck.

(4) Write an **A** next to the sentence below that is action only.

( ) John rang the bell twice.

( ) Dr. White's nightcap was droopy.

( ) Dr. White shut the window, and was soon at the door.

( ) John stood by me and stroked my neck.

( ) John knocked at the door.

( ) The knock was like thunder.

( ) I was very hot.

( ) In a minute we had left John far behind.

Way to go!

Tables & Graphs

Level ★

Score

/100

Math
DAY

30

Date   /   /

Name

1 The table and the graph pictured here both show the temperature over one   25 points per question
day. Answer the questions about the graph below.

Temperatures Throughout One Day

Time (o'clock)	6	7	8	9	10	11	12	1	2	3	4	5	6
Temperature (℃)	12	14	15	16	18	20	21	22	24	23	20	19	17

(1) Write the appropriate label for the horizontal axis.

(2) Write the appropriate label for the vertical axis.

(3) Complete the line graph by placing each point and then connecting them with a line.

(4) Write the title in box A.

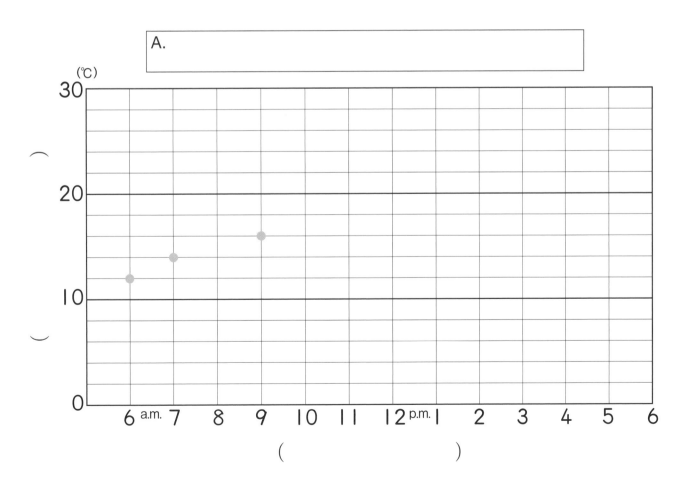

Reading
DAY
30

Actions & Descriptions
Black Beauty 4

Level ★★
Score

Date
/   /

Name

1 Read the passage. Then answer the questions below.

20 point
per q

    I will not tell about our way back. The doctor was a heavier man than John and not so good
a rider; however, I did my very best. The man at the toll-gate had it open. When we came to the
hill the doctor drew me up. "Now, my good fellow," he said, "take some breath." I was glad he
did, for I was nearly spent, but that breathing helped me on, and soon we were in the park. Joe
was at the lodge gate; my master was at the hall door, for he had heard us coming. He spoke
not a word; the doctor went into the house with him, and Joe led me to the stable. I was glad
to get home; my legs shook under me, and I could only stand and pant. I had not a dry hair
on my body, the water ran down my legs, and I steamed all over, Joe used to say, like a pot on
the fire. Poor Joe! He was young and small, and as yet he knew very little, and his father, who
would have helped him, had been sent to the next village; but I am sure he did the very best he
knew. He rubbed my legs and my chest, but he did not put my warm cloth on me; he thought
I was so hot I should not like it. Then he gave me a pailful of water to drink; it was cold and
very good, and I drank it all; then he gave me some hay and some corn, and thinking he had
done right, he went away. Soon I began to shake and tremble, and turned deadly cold; my legs
ached, my loins ached, and my chest ached, and I felt sore all over. Oh! How I wished for my
warm, thick cloth, as I stood and trembled. I wished for John, but he had eight miles to walk,
so I lay down in my straw and tried to go to sleep.

（1） How was the doctor different from John?

The doctor was a _____ man than John and not as _____ a rider.

（2） When they came to the hill, what did the doctor make Black Beauty do?

When they came to the hill, the doctor made Black Beauty take a _____.

（3） Describe Joe.

Joe was _____ and _____ and knew _____ about caring for horses.

（4） Write an **A** next to the sentences below that are actions only.
    (   ) I was like a pot on fire.
    (   ) I began to shake and tremble.
    (   ) I lay down in my straw.

（5） Write a **D** next to the sentences below that are descriptions only.
    (   ) I was glad to get home.
    (   ) I had not a dry hair on my body.
    (   ) He rubbed my legs and my chest.

Volume

Date   /   /

Name

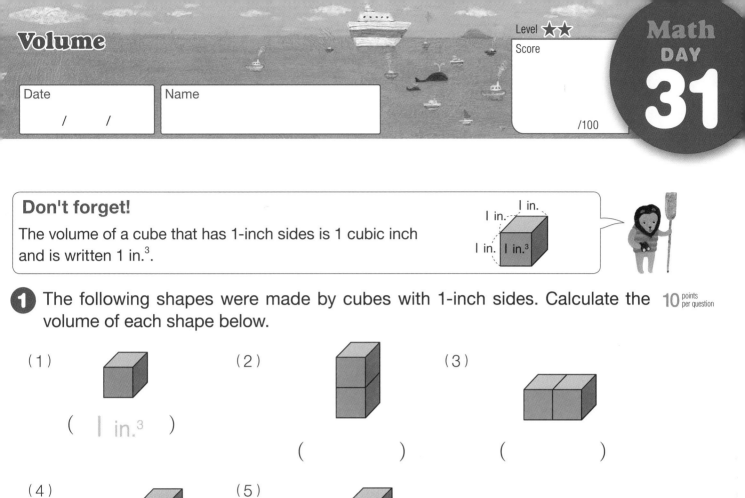

Don't forget!

The volume of a cube that has 1-inch sides is 1 cubic inch and is written 1 in.³

I in.
I in.
I in.   I in.³

1 The following shapes were made by cubes with 1-inch sides. Calculate the volume of each shape below.   10 points per question

(1)   (I in.³   )

(2)   (   )

(3)   (   )

(4)   (   )

(5)   (   )

2 Calculate the volume of the following rectangular prisms. Answer in cubic inches.   10 points per question

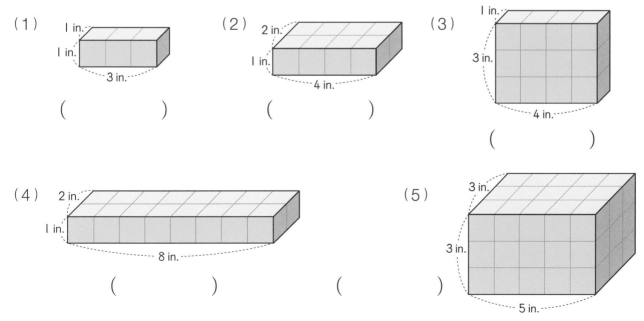

(1) I in. / I in. / 3 in.   (   )

(2) 2 in. / I in. / 4 in.   (   )

(3) I in. / 3 in. / 4 in.   (   )

(4) 2 in. / I in. / 8 in.   (   )

(5) 3 in. / 3 in. / 5 in.   (   )

Reading Comprehension
My Father's Dragon 1

Date   /   /

Name

① Read the excerpt from *My Father's Dragon* by Ruth Stiles Gannett. Then answer the questions below. 20 point per q

> The river was very wide and muddy, and the jungle was very gloomy and dense. The trees grew close to each other, and what room there was between them was taken up by great high ferns with sticky leaves. My father hated to leave the beach, but he decided to start along the river bank where at least the jungle wasn't quite so thick. He ate three tangerines, making sure to keep all the peels this time, and put on his rubber boots.
>
> My father tried to follow the river bank but it was very swampy, and as he went farther the swamp became deeper. When it was almost as deep as his boot tops he got stuck in the oozy, mucky mud. My father tugged and tugged, and nearly pulled his boots right off, but at last he managed to wade to a drier place. Here the jungle was so thick that he could hardly see where the river was. He unpacked his compass and figured out the direction he should walk in order to stay near the river. But he didn't know that the river made a very sharp curve away from him just a little way beyond, and so as he walked straight ahead he was getting farther and farther away from the river.

(1) Put a check (✓) next to the phrases that describe the story's setting.

( ) A dry and hot jungle   ( ) A wide and muddy river

( ) A gloomy and dense jungle   ( ) A river bank

( ) The lobby of a bank

(2) Who is the story about?

The story is about someone's _____.

(3) Who is telling the story?

The man's _____ or _____ is telling the story.

(4) Why does the father start along the river bank?

He starts along the river bank because the _____ isn't quite so _____ there.

(5) What does the narrator know that the father does not know?

The narrator knows that the river makes a very _____ and that

the father was getting _____ from the river.

> **Don't forget!** The **narrator** is the person who tells the events of a story. The **setting** is the background (specifically the time and place) of a story.

Volume

Date
/   /

Name

Don't forget!

The volume of a cube that has 1-centimeter sides is
1 cubic centimeter and is written 1 cm³.

I cm
I cm
I cm I cm³

1 The following shapes were made by cubes with 1-cm sides. Calculate the volume of each shape below. 10 points per question

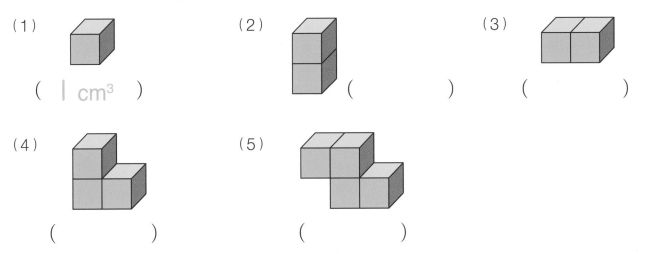

(1)

(I cm³)

(2)

( )

(3)

( )

(4)

( )

(5)

( )

2 Calculate the volume of the following rectangular prisms. Answer in cubic centimeters. 10 points per question

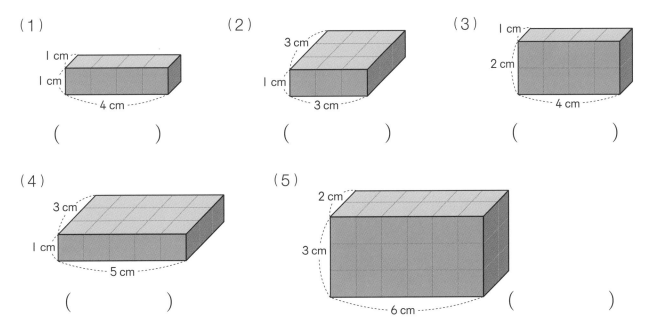

(1)

I cm
I cm
4 cm

( )

(2)

3 cm
I cm
3 cm

( )

(3)

I cm
2 cm
4 cm

( )

(4)

3 cm
I cm
5 cm

( )

(5)

2 cm
3 cm
6 cm

( )

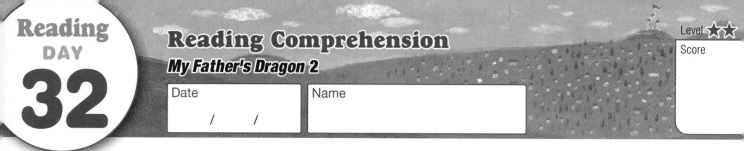

Reading DAY
32

Reading Comprehension
My Father's Dragon 2

Date / /

Name

Level ★★

Score

① Read the passage. Then answer the questions below.

> It was very hard to walk in the jungle. The sticky leaves of the ferns caught at my father's hair, and he kept tripping over roots and rotten logs. Sometimes the trees were clumped so closely together that he couldn't squeeze between them and had to walk a long way around.
>
> He began to hear whispery noises, but he couldn't see any animals anywhere. The deeper into the jungle he went the surer he was that something was following him, and then he thought he heard whispery noises on both sides of him as well as behind. He tried to run, but he tripped over more roots, and the noises only came nearer. Once or twice he thought he heard something laughing at him.
>
> At last he came out into a clearing and ran right into the middle of it so that he could see anything that might try to attack him. Was he surprised when he looked and saw fourteen green eyes coming out of the jungle all around the clearing, and then the green eyes turned into seven tigers! The tigers walked around him in a big circle, looking hungrier all the time, and then they sat down and began to talk.
>
> "I suppose you thought we didn't know you were trespassing in our jungle!"

（1）Why was it hard for the father to walk in the jungle?   30 poi for

It was hard for the father to walk in the jungle because he kept tripping over
_____ **and** _____.

（2）Complete the chart below that shows the cause of each of the father's actions.   40 po fo

Cause	Effect
The trees were clumped close together.	The father had to _____ a long way around.
The father heard whispery noises all around him.	He tried to _____.
The father wanted to see anything that might try to attack him.	He _____ into the middle of a _____.

（3）What did the tigers think that the father was doing?   30 poir

The tigers thought that the father was _____ **in their jungle.**

Don't forget! The **plot** is the main events of a story that are connected by cause and effect.

Capacity

Level ★★

Score

/100

Math
DAY
33

Date   /   /

Name

Don't forget!

2 cups = 1 pint (pt.)   8 fluid ounces (fl. oz.) = 1 cup

1 Convert the measurements below.

5 points per question

(1) 1 pt. = _____ cups

(2) 2 pt. = _____ cups

(3) 3 pt. = _____ cups

(4) 2 cups = _____ pt

(5) 1 cup = _____ pt.

(6) 4 cups = _____ pt.

2 Convert the measurements below.

5 points per question

(1) 1 cup = _____ fl. oz.

(2) 2 cups = _____ fl. oz.

(3) 4 cups = _____ fl. oz.

(4) 8 fl. oz. = _____ cup

(5) 24 fl. oz. = _____ cups

(6) 40 fl. oz. = _____ cups

3 How much water is in each measuring cup? Answer in two different units.

10 points per question

(1)   (2)   (3)   (4)

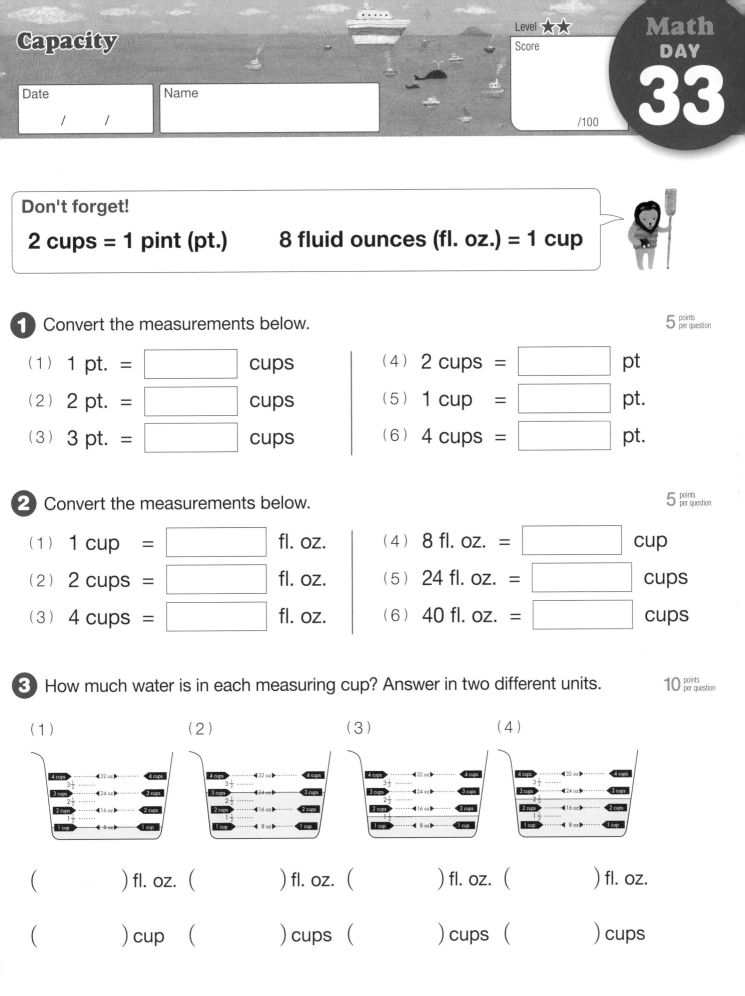

(   ) fl. oz. (   ) fl. oz. (   ) fl. oz. (   ) fl. oz.

(   ) cup   (   ) cups   (   ) cups   (   ) cups

Reading Comprehension
My Father's Dragon 3

Level ★★
Score

Date   /   /

Name

① Read the passage. Then answer the questions below.   100

Then the next tiger spoke. "I suppose you're going to say you didn't know it was our jungle!"

"Did you know that not one explorer has ever left this island alive?" said the third tiger.

My father thought of the cat* and knew this wasn't true. But of course he had too much sense to say so. One doesn't contradict a hungry tiger.

The tigers went on talking in turn. "You're our first little boy, you know. I'm curious to know if you're especially tender."

"Maybe you think we have regular meal-times, but we don't. We just eat whenever we're feeling hungry," said the fifth tiger.

"And we're very hungry right now. In fact, I can hardly wait," said the sixth.

"I can't wait!" said the seventh tiger.

And then all the tigers said together in a loud roar, "Let's begin right now!" and they moved in closer.

*In an earlier story in My Father's Dragon, the father meets a cat who was an explorer and had visited the island.

(1) Who is talking in this scene?

　　A group of _____ is talking in this scene.

(2) How many tigers are talking?

　　There are _____ tigers talking.

(3) What does the third tiger say?

　　The third tiger says that not one _____ has ever _____ the island alive.

(4) What does the fifth tiger say?

　　The fifth tiger says that they eat _____ they're feeling hungry.

(5) Does the father reply to the tigers?

　　_____, the father _____ reply to the tiger.

(6) What do the tigers say all together?

　　The tigers say in a loud roar, " _____ !"

Don't forget! **Dialogue** is the conversation between two or more characters in a story.

Date   /   /

Name

/100

> **Don't forget!**
> In order to find the area of a square or rectangle, use the following formulas:
> **A = l × w area of a rectangle = length × width**
> **A = s × s area of a square = side × side**

1 What is the area of each shape below? Answer in square inches and use the formulas from above.   8 points per question

(1) 1 in. / 1 in.   (   )

(2) 2 in. / 1 in.   (   )

(3) 2 in. / 2 in.   (   )

(4) 2 in. / 3 in.   (   )

(5) 3 in. / 8 in.   (   )

2 What is the area of each shape below? Answer each unit.   8 points per question

(1) 1 ft. / 2 ft.   (   )

(2) 2 ft. / 3 ft.   (   )

(3) 2 ft. / 5 ft.   (   )

(4) 3 cm / 4 cm   (   )

(5) 4 cm / 7 cm   (   )

3 What is the area of each shape below? Answer in square meters.   10 points per question

(1) 50 cm / 4 m   (   )

(2) 2 m / 1 m 50 cm   (   )

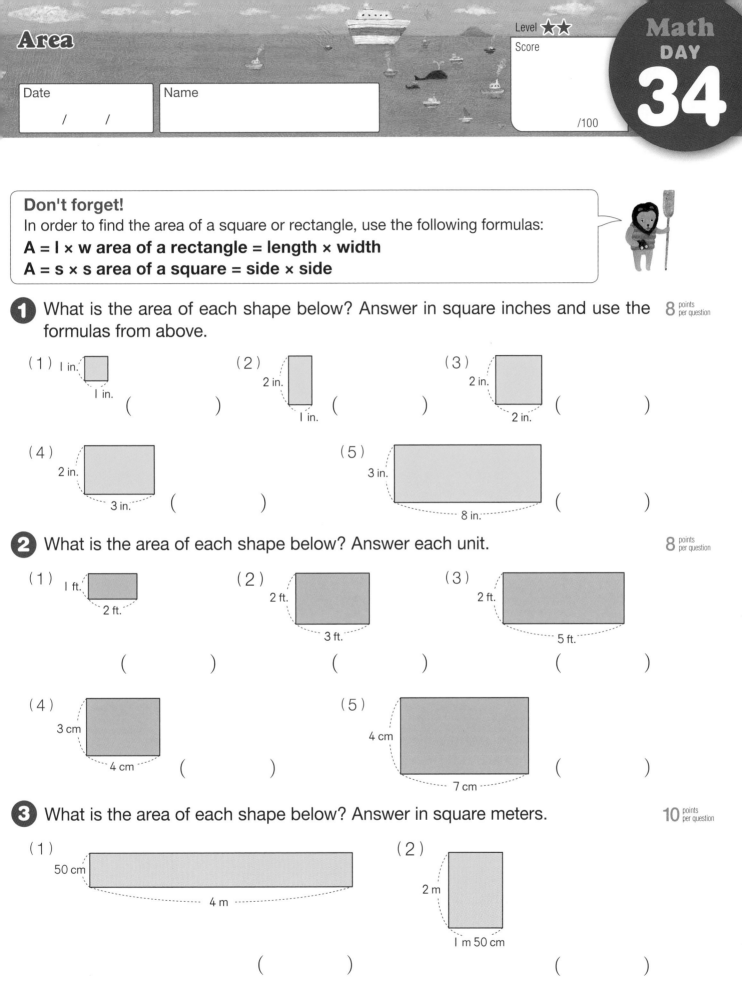

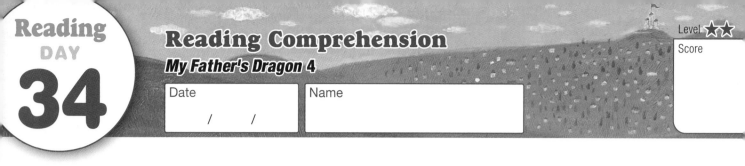

Reading
DAY
34

Reading Comprehension

My Father's Dragon 4

Date   /   /

Name

Level ★★

Score

① Read the passage. Then answer the questions below.

My father looked at those seven hungry tigers, and then he had an idea. He quickly opened his knapsack and took out the chewing gum. The cat had told him that tigers were especially fond of chewing gum, which was very scarce on the island. So he threw them each a piece but they only growled, "As fond as we are of chewing gum, we're sure we'd like you even better!" and they moved so close that he could feel them breathing on his face.

"But this is very special chewing gum," said my father. "If you keep on chewing it long enough it will turn green, and then if you plant it, it will grow more chewing gum, and the sooner you start chewing the sooner you'll have more."

The tigers said, "Why, you don't say! Isn't that fine!" And as each one wanted to be the first to plant the chewing gum, they all unwrapped their pieces and began chewing as hard as they could. Every once in a while one tiger would look into another's mouth and say, "Nope, it's not done yet," until finally they were all so busy looking into each other's mouths to make sure that no one was getting ahead that they forgot all about my father.

(1) Why did the father throw each tiger a piece of chewing gum?   25 point

The father threw each tiger a piece of chewing gum because the cat had told

him that tigers are especially _____.

(2) What was supposedly special about the chewing gum?   25 point for c

The chewing gum was supposedly special because if you kept chewing it, it

would turn_____ and you could _____ it to grow more gum.

(3) Complete the chart below that shows the effect of the father's action.   50 point for c

Cause	Effect
The father tells the tigers that the chewing gum has special powers.	All the tigers _____ their gum and began _____ it as hard as they could.
	The tigers would _____ to check if the gum was green.
	The tigers forgot all about the narrator's _____.

Elasped Time

Level ★★

Score

/100

Math
DAY
35

Date   /   /

Name

1 Use the figure above in order to answer the questions below.

10 points per question

(1) The short hand moves all the way around the clock once from ⬚ to noon.

(2) The short hand moves all the way around the clock once from noon to ⬚.

(3) The short hand moves all the way around the clock once every ⬚ hour(s).

(4) The short hand moves all the way around the clock ⬚ times in one day.

(5) ⬚ hours pass between midnight and 8 a.m.

(6) 9 hours after midnight, the time is ⬚ a.m.

2 Answer the questions below using the clock pictured here.

10 points per question

(1) Half an hour ago   (   )

(2) Half an hour from now   (   )

(3) An hour ago   (   )

(4) An hour from now   (   )

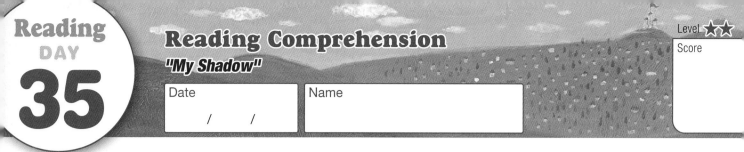

Reading Comprehension
"My Shadow"

Date   /   /

Name

1 Read the poem "My Shadow" by Robert Louis Stevenson. Then answer the 25 points per q
questions below.

I have a little shadow that goes in and out with me,
And what can be the use of him is more than I can see.
He is very, very like me from the heels up to the head;
And I see him jump before me, when I jump into my bed.

The funniest thing about him is the way he likes to grow—
Not at all like proper children, which is always very slow;
For he sometimes shoots up taller like an india-rubber ball,
And he sometimes gets so little that there's none of him at all.

He hasn't got a notion of how children ought to play,
And can only make a fool of me in every sort of way.
He stays so close beside me, he's a coward, you can see;
I'd think shame to stick to nursie as that shadow sticks to me!

One morning, very early, before the sun was up,
I rose and found the shining dew on every buttercup;
But my lazy little shadow, like an arrant* sleepy-head,
Had stayed at home behind me and was fast asleep in bed.
*"Arrant" means complete or utter.

(1) Which words in the first stanza rhyme with each other?

_____Me_____ rhymes with _____, and _____ rhymes with _____.

(2) Which words in the second stanza rhyme with each other?

_____ rhymes with _____, and _____ rhymes with _____.

(3) Put a check (✓) next to the phrases that describe the shadow in the poem.

( ) a copy-cat   ( ) grows and shrinks slowly
( ) shaped like the person   ( ) stays close
( ) always wandering away   ( ) sleeps all the time

(4) Put a check (✓) next to the phrases that describe some of the poem's main ideas.

( ) shadows follow you   ( ) shadows are like the sun
( ) shadows are lazy   ( ) shadows are good at games
( ) shadows change   ( ) your shadow is like you

Don't forget! A **stanza** is a group of lines that usually have a repeating pattern of rhythm and rhyme.

Elapsed Time

Level ★★

Score

/100

Math
DAY
36

1 How much time has passed from the time on the left to the time on the right? **10** points per question

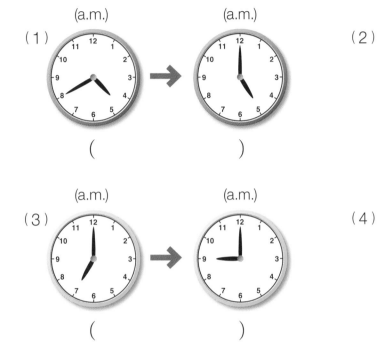

(1) (a.m.) → (a.m.)

( )

(2) (p.m.) → (p.m.)

( )

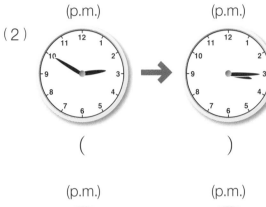

(3) (a.m.) → (a.m.)

( )

(4) (p.m.) → (p.m.)

( )

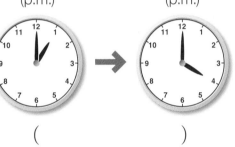

2 How much time has passed from the time on the left to the time on the right? **15** points per question

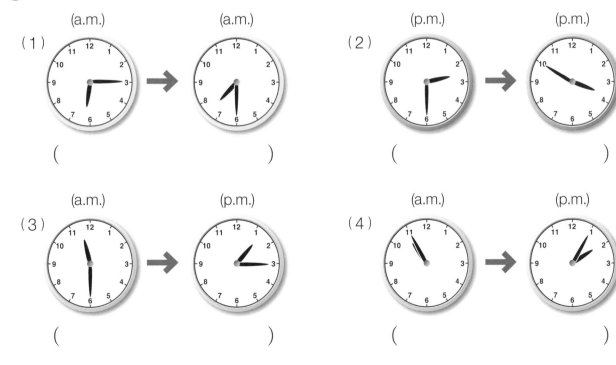

(1) (a.m.) → (a.m.)

( )

(2) (p.m.) → (p.m.)

( )

(3) (a.m.) → (p.m.)

( )

(4) (a.m.) → (p.m.)

( )

Reading Comprehension
"The Brook"

Date   /   /

Name

1 Read the excerpt from the poem "The Brook" by Alfred Tennyson. Then answer the questions below.

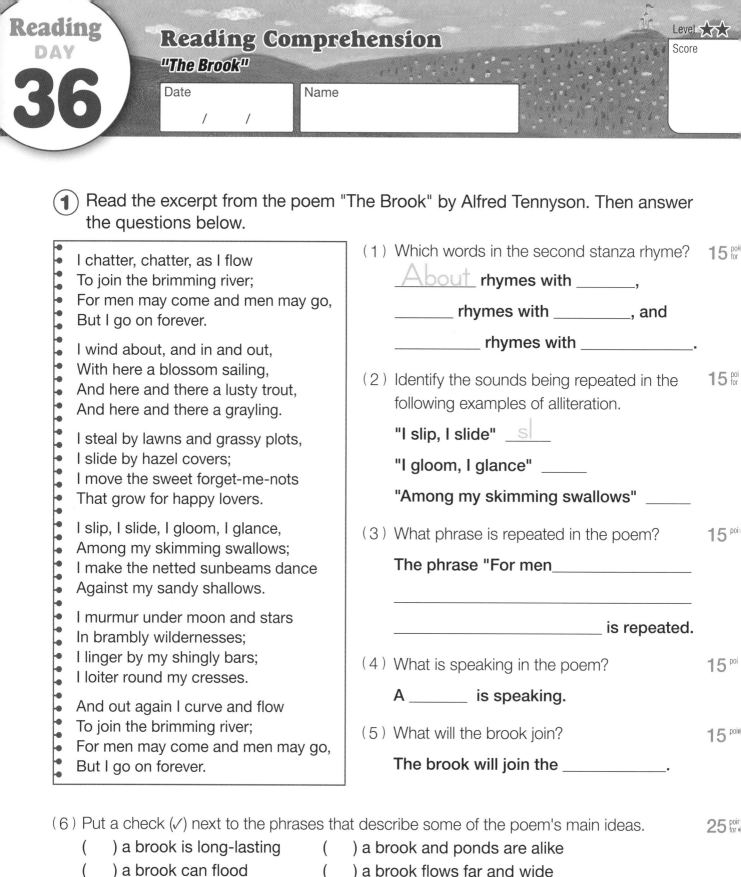

I chatter, chatter, as I flow
To join the brimming river;
For men may come and men may go,
But I go on forever.

I wind about, and in and out,
With here a blossom sailing,
And here and there a lusty trout,
And here and there a grayling.

I steal by lawns and grassy plots,
I slide by hazel covers;
I move the sweet forget-me-nots
That grow for happy lovers.

I slip, I slide, I gloom, I glance,
Among my skimming swallows;
I make the netted sunbeams dance
Against my sandy shallows.

I murmur under moon and stars
In brambly wildernesses;
I linger by my shingly bars;
I loiter round my cresses.

And out again I curve and flow
To join the brimming river;
For men may come and men may go,
But I go on forever.

(1) Which words in the second stanza rhyme?   15 poi for

About rhymes with _____,

_____ rhymes with _____, and

_____ rhymes with _____.

(2) Identify the sounds being repeated in the following examples of alliteration.   15 poi for

"I slip, I slide" __s|__

"I gloom, I glance" _____

"Among my skimming swallows" _____

(3) What phrase is repeated in the poem?   15 poi

The phrase "For men_____

_____

_____ is repeated.

(4) What is speaking in the poem?   15 poi

A _____ is speaking.

(5) What will the brook join?   15 poi

The brook will join the _____.

(6) Put a check (✓) next to the phrases that describe some of the poem's main ideas.   25 poi for

(   ) a brook is long-lasting   (   ) a brook and ponds are alike
(   ) a brook can flood   (   ) a brook flows far and wide
(   ) a brook is lively   (   ) a brook dries up

Don't forget! **Alliteration** is the repetition of a sound at the beginning of two or more words close to each other.

Circles & Spheres

Date   /   /

Name

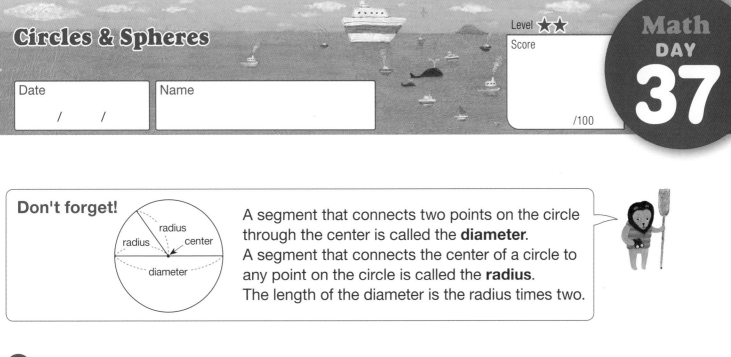

Don't forget!

radius
radius   center
diameter

A segment that connects two points on the circle through the center is called the **diameter**.
A segment that connects the center of a circle to any point on the circle is called the **radius**.
The length of the diameter is the radius times two.

1 Write the appropriate number in each box below.   15 points per question

(1) If the radius of a circle is 2 centimeters, the diameter is [   ] centimeters.

(2) If the diameter of a circle is 2 centimeters, the radius is [   ] centimeter(s).

(3) If the radius of a circle is 6 centimeters, the diameter is [   ] centimeters.

(4) If the diameter of a circle is 6 centimeters, the radius is [   ] centimeters.

2 Use a compass to draw a circle in each box below.   20 points per question

(1) Draw a circle with a radius of 2 centimeters.

(2) Draw a circle with a diameter of 5 centimeters.

Reading
DAY
37

Reading Comprehension
The Story of Doctor Dolittle 1

Level ★★
Score

Date   /   /

Name

1 Read the excerpt from *The Story of Doctor Dolittle* by Hugh Lofting. Then answer the questions below.

100

> Once upon a time, many years ago when our grandfathers were little children—there was a doctor; and his name was Dolittle—John Dolittle, M.D. "M.D." means that he was a proper doctor and knew a whole lot.
>
> He lived in a little town called Puddleby-on-the-Marsh. All the folks, young and old, knew him well by sight. And whenever he walked down the street in his high hat everyone would say, "There goes the Doctor!—He's a clever man." And the dogs and the children would all run up and follow behind him; and even the crows that lived in the church-tower would caw and nod their heads.
>
> The house he lived in, on the edge of the town, was quite small; but his garden was very large and had a wide lawn and stone seats and weeping-willows hanging over. His sister, Sarah Dolittle, was housekeeper for him; but the Doctor looked after the garden himself.
>
> He was very fond of animals and kept many kinds of pets. Besides the goldfish in the pond at the bottom of his garden, he had rabbits in the pantry, white mice in his piano, a squirrel in the linen closet and a hedgehog in the cellar. He had a cow with a calf too, and an old lame horse—twenty-five years of age—and chickens, and pigeons, and two lambs, and many other animals. But his favorite pets were Dab-Dab the duck, Jip the dog, Gub-Gub the baby pig, Polynesia the parrot, and the owl Too-Too.

(1) When does the story take place?

The story takes place many _____ ago.

(2) Who is the main character?

_____ is the main character.

(3) What is the setting of the story?

The setting of the story is the _____ of Dr. Dolittle at the edge of a little town

called _____ .

(4) What does Dr. Dolittle's sister, Sarah Dolittle, do?

Sarah Dolittle is Dr. Dolittle's _____.

(5) Why does Dr. Dolittle keep so many pets?

Dr. Dolittle keeps so many pets because he is very _____.

(6) Who are Dr. Dolittle's favorite pets?

Dr. Dolittle's favorite pets are _____, _____,

_____, _____ _____, and

_____ .

(7) Why does everyone think Dr. Dolittle is a clever man?

Everyone thinks Dr. Dolittle is a clever man because he is a _____.

Date   /   /

Name

1 Two circles of the same size are inside a larger circle that has a radius of 3 inches.

10 points per question

(1) What is the diameter of the big circle?   (   )

(2) What is the diameter of each small circle?   (   )

(3) What is the radius of each small circle?   (   )

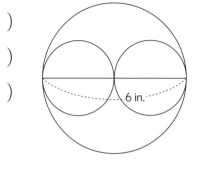

6 in.

2 The radius of the smallest circle in the figure below is 3 centimeters. How long is each side of the square?

35 points

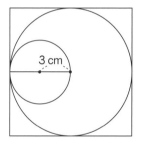

3 cm

(   )

3 Each circle in the figure below has a radius of 4 inches. How long is the line from A to B?

35 points

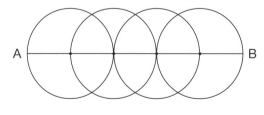

A   B

(   )

Reading
DAY
38

Reading Comprehension
The Story of Doctor Dolittle 2

Level ★★
Score

Date / /

Name

1 Read the passage. Then answer the questions below.

25 points per qu.

His sister used to grumble about all these animals and said they made the house untidy. And one day when an old lady with rheumatism came to see the Doctor, she sat on the hedgehog who was sleeping on the sofa and never came to see him anymore, but drove every Saturday all the way to Oxenthorpe, another town ten miles off, to see a different doctor.

Then his sister, Sarah Dolittle, came to him and said, "John, how can you expect sick people to come and see you when you keep all these animals in the house? It's a fine doctor would have his parlor full of hedgehogs and mice! That's the fourth personage these animals have driven away. Squire Jenkins and the Parson say they wouldn't come near your house again—no matter how sick they are. We are getting poorer every day. If you go on like this, none of the best people will have you for a doctor."

"But I like the animals better than the 'best people'," said the Doctor.

"You are ridiculous," said his sister, and walked out of the room.

So, as time went on, the Doctor got more and more animals; and the people who came to see him got less and less. Till at last he had no one left—except the Cat's-meat-Man, who didn't mind any kind of animal. But the Cat's-meat-Man wasn't very rich and he only got sick once a year—at Christmas-time, when he used to give the Doctor sixpence for a bottle of medicine.

(1) Why did Sarah grumble about all the Doctor's animals?

Sarah grumbled about the animals because she said they_____.

(2) Why did the old lady with rheumatism never come back to the Doctor?

The old lady never came back because she accidentally _____.

(3) Complete the chart with words from the passage above.

Cause	Effect
Because the old lady sat on the hedgehog.	She drove all the way to _____ to see a different _____.
Because of the Doctor's pets	Less and less _____ _____.
Because the Cat's-meat-Man didn't mind _____.	He would still visit Doctor Dolittle when he was sick.

(4) What is the main idea of this passage?

The main idea of the passage is: Because Doctor Dolittle got more and more

_____, **he got** _____ **patients.**

Circles & Spheres

Level ★★

Score

/100

Math
DAY
39

Don't forget!

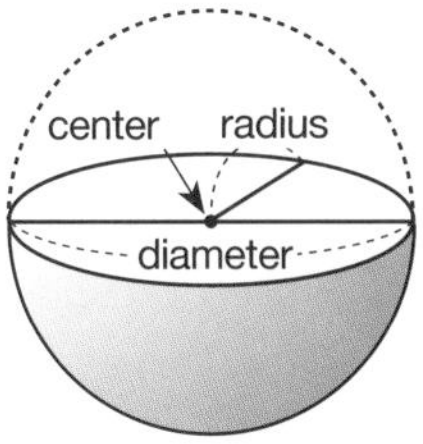

center radius

diameter

If you cut a sphere in half, the center, radius and diameter of the circle in its cross section is equal to the center, radius and diameter of the sphere.

1 Write the appropriate number in each box below.

10 points per question

(1) If the diameter of a sphere is 6 centimeters, the radius is [ ] centimeters.

(2) If the diameter of a sphere is 5 centimeters,

its radius is [ ] centimeters and [ ] millimeters.

(3) If the radius of a sphere is 4 centimeters, the diameter is [ ] centimeters.

(4) If the radius of a sphere is 5 centimeters, the diameter is [ ] centimeters.

2 As pictured on the right, you have a sphere that fits snugly inside a box.

12 points per question

(1) How long is each side of the box? (   )

(2) How long is the diameter of the sphere? (   )

(3) How long is the radius of the sphere? (   )

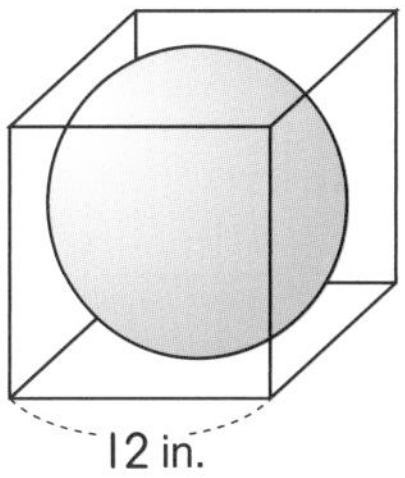

12 in.

3 As pictured on the right, you have six balls that fit snugly inside one box that is 15 inches wide.

12 points per question

(1) What is the diameter of each ball? (   )

(2) What is the length of the box? (   )

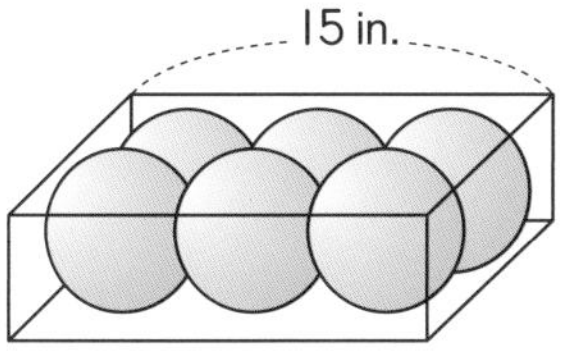

15 in.

Reading
DAY
39

Level ⭐⭐
Score

Reading Comprehension
The Story of Doctor Dolittle 3

Date   /   /

Name

① Read the passage. Then answer the questions below.

Sixpence a year wasn't enough to live on—even in those days, long ago; and if the Doctor hadn't had some money saved up in his money-box, no one knows what would have happened.

And he kept on getting still more pets; and of course it cost a lot to feed them. And the money he had saved up grew littler and littler.

Then he sold his piano, and let the mice live in a bureau-drawer. But the money he got for that too began to go, so he sold the brown suit he wore on Sundays and went on becoming poorer and poorer.

And now, when he walked down the street in his high hat, people would say to one another, "There goes John Dolittle, M.D.! There was a time when he was the best known doctor in the West Country—Look at him now—He hasn't any money and his stockings are full of holes!"

But the dogs and the cats and the children still ran up and followed him through the town—the same as they had done when he was rich.

(1) Complete the chart with words from the passage above.   60 poi for

Cause	Effect
Sixpence a year wasn't enough to live on.	The doctor spent his _____ money.
He kept on getting still more pets.	The doctor sold _____.
The money from his piano wasn't enough.	The doctor sold his _____and became _____.

(2) Identify the statements as **T** (true) or **F** (false) according to the passage.   10 poir per

(1) The mice began living in the Doctor's bureau-drawer.   **T   F**

(2) Sixpence a year was just enough to live on.   **T   F**

(3) The Doctor kept losing his pets.   **T   F**

(3) Who didn't care that Doctor Dolittle didn't have any money?   10 poir for

The _____ and the _____ and the _____ didn't care that Doctor Dolittle didn't have any money.

Triangles

Level ★★

Score

/100

Math
DAY

40

Date   /   /

Name

Don't forget!

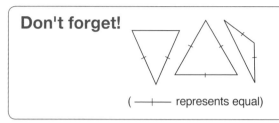

(—+— represents equal)

Congruent means equal in size and shape.
A triangle with two congruent sides is an
isosceles triangle.
A triangle with three congruent sides is
called an **equilateral** triangle.

1 Sort the triangles below into equilateral and isosceles triangles by putting
each letter next to the correct category.

20 points
per question

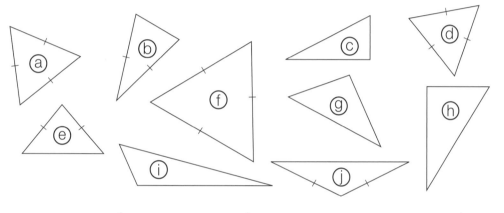

(1) equilateral triangle (   )   (2) isosceles triangle (   )

2 Use your ruler and compass to draw the triangles below.

30 points
per question

(1) Draw a triangle with sides that are 4 centimeters, 2 centimeters, and 4 centimeters
long.

(2) Draw a triangle with sides that are all 3 centimeters long.

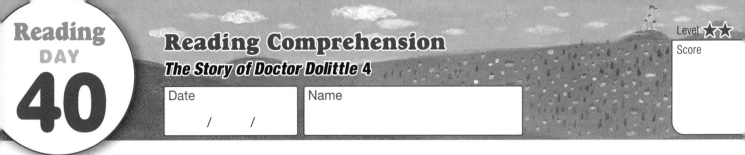

Reading
DAY
40

Reading Comprehension
The Story of Doctor Dolittle 4

Date   /   /

Name

Level ★★

Score

① Read the passage. Then answer the questions below.

20 points per qu

> It happened one day that the Doctor was sitting in his kitchen talking with the Cat's-meat-Man who had come to see him with a stomach-ache.
>
> "Why don't you give up being a people's doctor, and be an animal-doctor?" asked the Cat's-meat-Man.
>
> The parrot, Polynesia, was sitting in the window looking out at the rain and singing a sailor-song to herself. She stopped singing and started to listen.
>
> "You see, Doctor," the Cat's-meat-Man went on, "you know all about animals—much more than what these here vets do. That book you wrote—about cats, why, it's wonderful! I can't read or write myself—or maybe I'd write some books. But my wife, Theodosia, she's a scholar, she is. And she read your book to me. Well, it's wonderful—that's all can be said—wonderful. You might have been a cat yourself. You know the way they think. And listen: you can make a lot of money doctoring animals. Do you know that? You see, I'd send all the old women who had sick cats or dogs to you. And if they didn't get sick fast enough, I could put something in the meat I sell 'em to make 'em sick, see?"
>
> "Oh, no," said the Doctor quickly. "You mustn't do that. That wouldn't be right."
>
> "Oh, I didn't mean real sick," answered the Cat's-meat-Man. "Just a little something to make them droopy-like was what I had reference to. But as you say, maybe it ain't quite fair on the animals. But they'll get sick anyway, because the old women always give 'em too much to eat. And look, all the farmers 'round about who had lame horses and weak lambs— they'd come. Be an animal-doctor."

(1) Who is speaking in this passage?

The _____ and the _____ are speaking in this passage.

(2) Who is listening in this scene?

The _____ is listening.

(3) What is the Cat's-meat-Man's idea?

The Cat's-meat-Man's idea is that the Doctor should become an_____.

(4) What are some supporting details for Cat's-meat-Man's suggestion?

(a) The Doctor knows all about _____.

(b) The Doctor wrote a wonderful _____ about _____.

(c) The Doctor can make a lot of _____ doctoring animals.

(5) What does the Doctor tell the Cat's-meat-Man not to do?

The Doctor tells him not to put something in the _____ to make the

animals _____.

Reading is fun!

Angles

Date   /   /

Name

Don't forget!
An **angle** is the geometric figure formed by two distinct rays that have one common endpoint. This common endpoint is called the **vertex** and the rays are called the **sides** of the angle. The measure of an angle is the size of the space between those two distinct rays that have one common endpoint. Angles are measured in degrees.

side
angle
side

1 Use the protractors to measure each angle below.

10 points per question

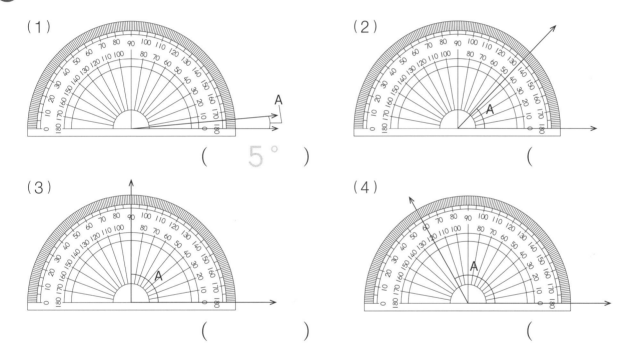

(1)

(5°   )

(2)

(   )

(3)

(   )

(4)

(   )

2 Use your own protractor to measure the angles below.

15 points per question

(1)

A

(   )

(2)

A

(   )

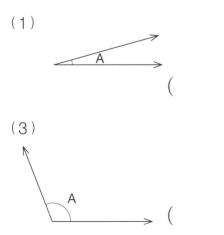

(3)

A

(   )

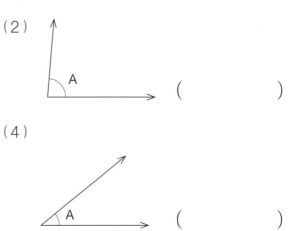

(4)

A

(   )

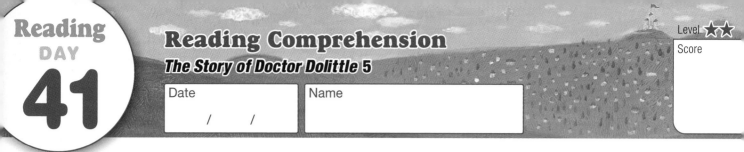

Reading
DAY
41

Reading Comprehension
The Story of Doctor Dolittle 5

Level ★★

Score

Date / /

Name

① Read the passage. Then answer the questions below.

20 points per q

When the Cat's-meat-Man had gone the parrot flew off the window on to the Doctor's table and said, "That man's got sense. That's what you ought to do. Be an animal-doctor. Give the silly people up—if they haven't brains enough to see you're the best doctor in the world. Take care of animals instead—They'll soon find it out. Be an animal-doctor."

"Oh, there are plenty of animal-doctors," said John Dolittle, putting the flower-pots outside on the window sill to get the rain.

"Yes, there ARE plenty," said Polynesia. "But none of them are any good at all. Now listen, Doctor, and I'll tell you something. Did you know that animals can talk?"

"I knew that parrots can talk," said the Doctor.

"Oh, we parrots can talk in two languages—people's language and bird-language," said Polynesia proudly. "If I say, 'Polly wants a cracker,' you understand me. But hear this: Ka-ka oi-ee, fee-fee?"

"Good gracious!" cried the Doctor. "What does that mean?"

"That means, 'Is the porridge hot yet?'—in bird-language."

"My! You don't say so!" said the Doctor. "You never talked that way to me before."

"What would have been the good?" said Polynesia, dusting some cracker-crumbs off her left wing. "You wouldn't have understood me if I had."

"Tell me some more," said the Doctor, all excited; and he rushed over to the dresser-drawer and came back with the butcher's book and a pencil. "Now don't go too fast—and I'll write it down. This is interesting—very interesting—something quite new. Give me the Birds' ABCs first—slowly now."

(1) Who is speaking this passage?

The _____ and the _____ are speaking in the passage.

(2) What languages does Polynesia speak?

Polynesia speaks _____ and _____.

(3) Why didn't Polynesia talk to the Doctor in bird-language before?

Polynesia didn't talk to the Doctor in bird-language before because he

_____.

(4) Why is the Doctor excited?

The Doctor is excited because he is going to learn _____.

(5) How does the Doctor start to learn bird-language?

The Doctor starts by learning the _____ first.

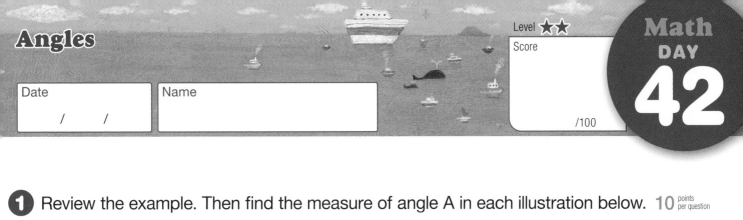

Angles

Level ★★

Score

/100

Math
DAY
42

Date / /

Name

---

1 Review the example. Then find the measure of angle A in each illustration below. **10** points per question

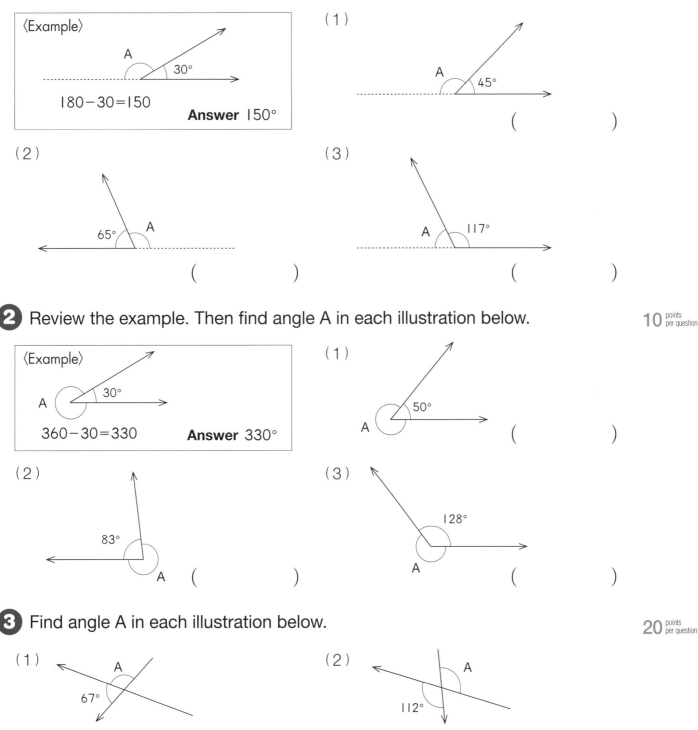

〈Example〉

A 30°

180−30=150

Answer 150°

(1)

A 45°

(   )

(2)

65° A

(   )

(3)

A 117°

(   )

2 Review the example. Then find angle A in each illustration below. **10** points per question

〈Example〉

A 30°

360−30=330   **Answer** 330°

(1)

50° A

(   )

(2)

83° A

(   )

(3)

128° A

(   )

3 Find angle A in each illustration below. **20** points per question

(1)

A 67°

(   )

(2)

A 112°

(   )

umon Publishing Co.,Ltd.

83

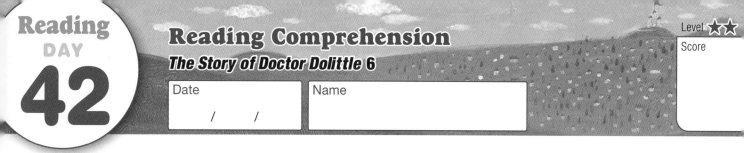

Reading Comprehension
The Story of Doctor Dolittle 6

Date   /   /   Name

① Read the passage. Then answer the questions below.   20 poin per •

So that was the way the Doctor came to know that animals had a language of their own and could talk to one another. And all that afternoon, while it was raining, Polynesia sat on the kitchen table giving him bird words to put down in the book.

At tea-time, when the dog, Jip, came in, the parrot said to the Doctor, "See, he's talking to you."

"Looks to me as though he were scratching his ear," said the Doctor.

"But animals don't always speak with their mouths," said the parrot in a high voice, raising her eyebrows. "They talk with their ears, with their feet, with their tails—with everything. Sometimes they don't WANT to make a noise. Do you see now the way he's twitching up one side of his nose?"

"What's that mean?" asked the Doctor.

"That means, 'Can't you see that it has stopped raining?'" Polynesia answered. "He is asking you a question. Dogs nearly always use their noses for asking questions."

After a while, with the parrot's help, the Doctor got to learn the language of the animals so well that he could talk to them himself and understand everything they said. Then he gave up being a people's doctor altogether.

As soon as the Cat's-meat-Man had told every one that John Dolittle was going to become an animal-doctor, old ladies began to bring him their pet pugs and poodles who had eaten too much cake; and farmers came many miles to show him sick cows and sheep.

（1） What did Doctor Dolittle write down in his book?

The Doctor wrote down _____ in his book.

（2） Who was talking to Doctor Dolittle with his nose?

The _____ was talking to the Doctor with his nose.

（3） How was the Doctor able to learn the language of the animals?

The Doctor was able to learn the language of the animals with the help of the

_____.

（4） Why don't animals always speak with their mouths?

Animals don't always speak with their mouths because sometimes they don't

want to make _____.

（5） Who brought their sick animals to Doctor Dolittle?

The _____ and the _____ brought their sick animals.

Date / /

Name

1 Multiply or divide.

5 points per question

(1)
```
   4 0
 ×   2
```

(5)
```
   3 2
 × 2 3
```

(9)
```
55 ) 3 7 5
```

(13)
```
29 ) 1 5 0 6
```

(2)
```
   2 7 3
 ×     3
```

(6)
```
   1 1 5
 × 1 3 4
```

(10)
```
8 ) 5 8 4 6
```

(14)
```
18 ) 5 5 0
```

(3)
```
   3 0 7
 ×     7
```

(7)
```
   2 3 0
 × 1 2 5
```

(11)
```
5 ) 8 0 6
```

(15)
```
328 ) 1 8 1 0
```

(4)
```
   2 0 1
 × 1 1 3
```

(8)
```
   1 6
 ×   7
```

(12)
```
123 ) 2 2 2
```

(16)
```
97 ) 7 3 2
```

2 Stamps are sold in rolls of 45. If Jane buys 12 rolls, how many stamps did she get? 10 points

Ans. _____

3 There are 196 roses, and the florist split them evenly into 14 bunches. How many roses are in each bunch? 10 points

Ans. _____

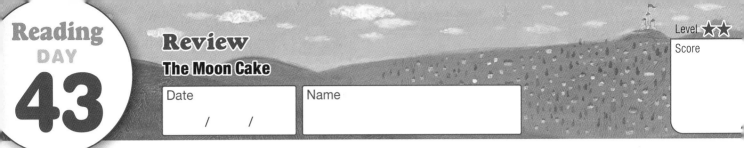

Review
The Moon Cake

Level ★★
Score

Date   /   /

Name

① Read the passage. Then answer the questions below.

A little monkey had a cake that a big monkey coveted. The big monkey made a plan to get the cake without making the little monkey cry so loud as to attract his mother's attention. The big monkey told the little monkey that the cake would be prettier if it were more like the moon. The big monkey thought that a cake like the moon must be beautiful, and on being assured by the big monkey that he had made many such moon cakes, he handed over his cake for manipulation. The big monkey took a big mouthful, leaving a crescent with jagged edges. The little monkey was not pleased by the change, and began to whimper; but the big monkey silenced him by saying that he would make the cake into a half-moon. So he nibbled off the horns of the crescent, and gnawed the edge smooth; but when the half-moon was made, the little monkey saw that there was hardly any cake left, and he again began to cry again. The big monkey again diverted him by telling him that, if he did not like so small a moon, he should have one that was just the size of the real thing. He then took the cake, and explained that, just before the new moon is seen, the old moon disappears. Then he swallowed the rest of the cake and ran away. And while little monkey waited for the new moon, the big monkey cried because his stomach hurt from too much cake.

（1）Choose words from the passage to complete the definitions below.   5 points per word

_____ wanted to have something

_____ more attractive

_____ convinced; got rid of any doubts

_____ changes

_____ whine; cry; sob

_____ distracted; sidetracked

（2）How did the big monkey trick the little monkey?   30 points for c

The big monkey convinced the little monkey that the cake would be _____

if it were more like the _____.

（3）Complete the chart with words from the passage above.   40 points for c

Cause	Effect
The little monkey had a cake that the big monkey wanted.	The big monkey _____ get the cake.
The big monkey took a big mouthful.	The little monkey began to _____.
The big monkey _____ the _____ the cake.	The big monkey's _____ hurt.

Level ★★

Score

/100

Date   /   /

Name

1 Calculate.   5 points per question

(1)
$$\begin{array}{r} 1.7 \\ +\ 2.1 \\ \hline \end{array}$$

(3)
$$\begin{array}{r} 12.5 \\ -\ 1.5 \\ \hline \end{array}$$

(5)
$$\begin{array}{r} 0.5 \\ +\ 10.7 \\ \hline \end{array}$$

(7)
$$\begin{array}{r} 1.07 \\ +\ 1.40 \\ \hline \end{array}$$

(2)
$$\begin{array}{r} 2.1 \\ -\ 0.4 \\ \hline \end{array}$$

(4)
$$\begin{array}{r} 3.7 \\ +\ 11.5 \\ \hline \end{array}$$

(6)
$$\begin{array}{r} 12.5 \\ -\ 8.0 \\ \hline \end{array}$$

(8)
$$\begin{array}{r} 2.30 \\ -\ 0.68 \\ \hline \end{array}$$

2 Calculate.   5 points per question

(1) $\dfrac{3}{5} + \dfrac{1}{5} =$

(3) $\dfrac{4}{9} + \dfrac{3}{9} =$

(5) $\dfrac{6}{7} - \dfrac{3}{7} =$

(2) $\dfrac{2}{7} + \dfrac{3}{7} =$

(4) $\dfrac{4}{5} - \dfrac{1}{5} =$

(6) $\dfrac{7}{9} - \dfrac{5}{9} =$

3 Cindy's bag weighs 2.5 kilograms. Her sister's bag is 700 grams heavier. How 15 points
much does her sister's bag weigh? Answer in kilograms.

Ans. _____

4 Fred's living room is 3 meters and 45 centimeters wide. His pictures are 40 15 points
centimeters wide, and he wants the pictures spaced equally as shown below.
How much space should he put between the pictures in his living room?

40 cm

Ans. _____

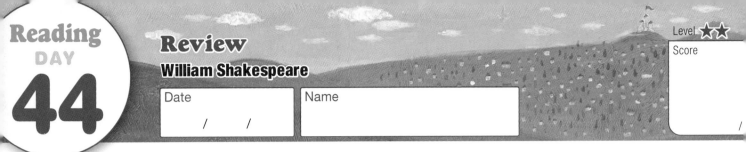

Reading DAY 44

Review
William Shakespeare

Level ★★

Score

Date / /

Name

/

1 Read the passage. Then answer the questions below.

100 points

> William Shakespeare is an author who has been entertaining readers and theater audiences for centuries. His plays and poems appeared in the late 16th century in England and are still read and performed worldwide today.
>
> It is unclear how Shakespeare's theater career began. Most of what we know about Shakespeare is from public records. It is known that Shakespeare arrived in London in his mid-twenties. He began modeling his own plays on the successful plays appearing in London theaters. Shakespeare had the most success with comedy—and particularly romantic comedies. Shakespeare also began including English history into his plays, which made his plays stand out. At that point in time, the historic play was a new genre (a category of art, such as in literature). Shakespeare blended comedy and tragedy to make the genre his own.
>
> From 1594 and on he was a part of a theater group called "Lord Chamberlain's Men." This group became a big hit at the Globe Theatre in London. The Globe Theatre had a unique design—the building was a circular shape and the audience would gather around the stage in a semi-circle. Shakespeare staged many of his plays specifically for the Globe Theatre.
>
> Around 1594, Shakespeare also began writing *Romeo and Juliet*, which would become one of his most famous plays. The play focuses on a young man and woman who fall in love but are torn apart because their families are enemies. *Romeo and Juliet* continues to be performed and adapted for stage, film, television and more. Indeed, Shakespeare's plays are as alive today as they were in 16th century England.

(1) Who was William Shakespeare?

William Shakespeare was an _____.

(2) What did Shakespeare write?

Shakespeare wrote _____ and _____.

(3) Where did Shakespeare stage many of his plays?

Shakespeare staged many of his plays at the _____ in _____.

(4) When did Shakespeare begin working with Lord Chamberlain's Men?

Shakespeare began working with Lord Chamberlain's Men in _____.

(5) How did Shakespeare stage his plays?

Shakespeare staged many of his plays specifically for the _____.

(6) Why did Shakespeare's plays about English history stand out?

Shakespeare's plays about English history stood out because they were a new

_____ , and he blended _____ and _____.

Review

Level ★★

Score

Math
DAY

45

/100

Date

/   /

Name

1 How much time has passed from the time on the left to the time on the right?   15 points per question

(a.m.)   (a.m.)   (a.m.)   (p.m.)

(1)

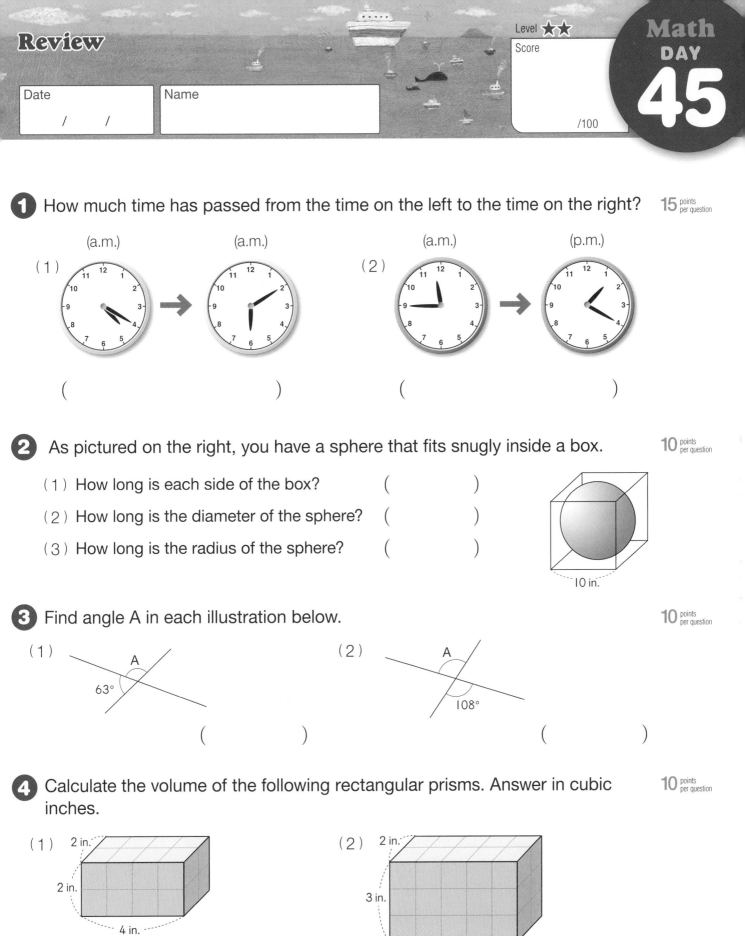

(   )   (   )

2 As pictured on the right, you have a sphere that fits snugly inside a box.   10 points per question

(1) How long is each side of the box?   (   )

(2) How long is the diameter of the sphere?   (   )

(3) How long is the radius of the sphere?   (   )

10 in.

3 Find angle A in each illustration below.   10 points per question

(1)   A

63°

(   )

(2)   A

108°

(   )

4 Calculate the volume of the following rectangular prisms. Answer in cubic inches.   10 points per question

(1) 2 in.

2 in.

4 in.

(   )

(2) 2 in.

3 in.

5 in.

(   )

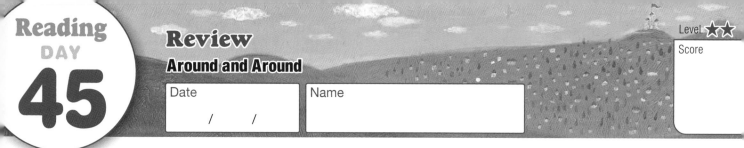

Reading
DAY
45

Review
Around and Around

Date   /   /

Name

Level ⭐⭐

Score

① Read the passage. Then answer the questions below.   25 points per q

Have you ever seen a professional cyclist ride upside-down in a loop? The cyclist is taking advantage of different forces at work: centrifugal force and centripetal force.

The same force is at work if a person swings a bucket of water. By attaching a small bucket of water to the end of a rope, a person can swing the bucket in a circle fast enough so that the water will not spill out. The water tends to stay in a straight line and so it pushes toward the bottom of the bucket. This is called centrifugal force—the tendency for any object to continue to move in a straight line and move away from the center. Centripetal force also is working at the same time. Centripetal force is the force that keeps an object moving along a curved path, so the force is directed inward toward the center. The rope is the centripetal force that keeps the bucket moving in a circle. Without the rope (or if you let go of the rope), the bucket would immediately fly off in a straight line.

Both centrifugal and centripetal force help the professional cyclist perform his trick. As the cyclist travels up the loop and upside-down at the top, he stays on the track because of centrifugal force. The cyclist and the bike continue to move in a straight line. Meanwhile, the track acts as the centripetal force that keeps the cyclist moving in a circular direction.

(1) Describe the way the bucket and the bike must be moving in order to do the trick.

The bucket and the bike must be moving _____ in order to do the trick.

(2) What can the professional cyclist do because of centrifugal and centripetal force?

The professional cyclist can ride _____ in a loop.

(3) What is the main idea of the whole passage?

The main idea is that a cyclist takes advantage of _____ force and

_____ force to ride upside-down in a loop.

(4) What details support the main idea? Put a check (✓) next to the answers.

(   ) The cyclist and the bike continue to move in a straight line.

(   ) The force of gravity holds objects down.

(   ) The track keeps the cyclist moving in a circular direction.

(   ) Many inventions use centrifugal and centripetal force.

90

Wow! You finished!
Congratulations!

DAY 1, pages 1 & 2

1 Multiply.

(1) 13 × 2 = 26
(2) 34 × 2 = 68
(3) 23 × 3 = 69
(4) 43 × 3 = 129
(5) 32 × 4 = 128
(6) 43 × 4 = 172
(7) 24 × 5 = 120
(8) 37 × 5 = 185
(9) 27 × 6 = 162
(10) 48 × 6 = 288
(11) 31 × 7 = 217
(12) 53 × 7 = 371
(13) 25 × 8 = 200
(14) 39 × 8 = 312
(15) 16 × 9 = 144
(16) 57 × 9 = 513
(17) 60 × 4 = 240
(18) 73 × 2 = 146
(19) 36 × 3 = 108
(20) 84 × 5 = 420

① Read each word aloud. Then divide the word into syllables.

(1) airplane — air/plane
(2) parachute — par/a/chute
(3) scrapbook — scrap/book
(4) automobile — au/to/mo/bile
(5) highway — high/way
(6) computer — com/put/er
(7) basketball — bas/ket/ball
(8) spectators — spec/ta/tors

② Write the words with the same amount of syllables in each group below.

alligator propeller hollow overboard pelican motorcycle dictionary badger hurdle

(a) 2 syllables: badger, hollow, hurdle
(b) 3 syllables: overboard, pelican, propeller
(c) 4 syllables: alligator, dictionary, motorcycle

DAY 2, pages 3 & 4

1 Multiply.

(1) 130 × 4 = 520
(2) 273 × 2 = 546
(3) 408 × 3 = 1224
(4) 318 × 3 = 954
(5) 116 × 4 = 464
(6) 347 × 4 = 1388
(7) 227 × 5 = 1135
(8) 547 × 5 = 2735
(9) 409 × 6 = 2454
(10) 647 × 6 = 3882
(11) 503 × 7 = 3521
(12) 381 × 7 = 2667
(13) 308 × 8 = 2464
(14) 459 × 8 = 3672
(15) 207 × 9 = 1863
(16) 728 × 9 = 6552
(17) 274 × 6 = 1644
(18) 686 × 7 = 4802
(19) 778 × 8 = 6224
(20) 889 × 9 = 8001

① Complete the table below according to the example.

adjective	adverb	adjective	adverb
immediate	immediately	lazy	lazily
deliberate	deliberately	sleepy	sleepily
normal	normally	merry	merrily
polite	politely	helpful	helpfully
rapid	rapidly	generous	generously
playful	playfully	light	lightly
swift	swiftly	dainty	daintily

② Read the passage. Then answer the questions below using only adverbs from the passage.

An old woman with a large bag boarded a bus that rapidly drove away. Immediately, a young man generously gave his seat to her. Normally, most people would swiftly take the seat and politely thank the man for being so helpful. Instead, the woman daintily laid her bag on the seat and remained standing. Was the woman deliberately being rude? No, she merrily explained that inside the bag was a litter of sleepy kittens. She was carefully taking them to a new home.

(1) How did the bus drive?
The bus drove rapidly.
(2) When did the young man give up his seat?
The young man gave up his seat immediately.
(3) How would people normally react to the young man's offer?
Normally, most people would politely thank the man.
(4) How did the woman lay her bag down?
The woman laid her bag down daintily.
(5) How was the woman taking the kittens to a new home?
The woman was carefully taking the kittens to a new home.

DAY 3, pages 5 & 6

1 Multiply.

(1) 32 × 12: 64, 32, 384
(2) 42 × 23: 126, 84, 966
(3) 43 × 34: 172, 129, 1462
(4) 54 × 37: 378, 162, 1998
(5) 34 × 45: 170, 136, 1530
(6) 46 × 42: 92, 184, 1932
(7) 23 × 52: 46, 115, 1196
(8) 38 × 55: 190, 190, 2090
(9) 47 × 60 = 2820
(10) 28 × 61: 28, 168, 1708
(11) 32 × 74: 128, 224, 2368
(12) 53 × 72: 106, 371, 3816
(13) 29 × 81: 29, 232, 2349
(14) 34 × 83: 102, 272, 2822
(15) 26 × 93: 78, 234, 2418
(16) 54 × 91: 54, 486, 4914
(17) 50 × 41: 50, 200, 2050
(18) 67 × 24: 268, 134, 1608
(19) 35 × 38: 280, 105, 1330
(20) 84 × 58: 672, 420, 4872

① Trace the words below.

(1) knead
(2) wrench
(3) kneel
(4) knuckles
(5) bomb
(6) pneumonia
(7) limb
(8) cologne
(9) wreckage

② Choose a word from the list above to complete each sentence.

(1) My dad always puts on cologne before he goes to work.
(2) The divers searched the wreckage for lost treasure.
(3) The mechanic asked his assistant for a wrench.
(4) The baker will knead the dough before rolling it out.
(5) After I practiced on the punching bag, my knuckles hurt.
(6) Her cousin went to the hospital because she had pneumonia.
(7) The movie was a box-office bomb. It got terrible reviews.
(8) The dog broke his leg, so the doctors put a cast on the limb.
(9) When we finally got to the cave, we had to kneel because the ceiling was so low.

DAY 4, pages 7 & 8

1 Multiply.

(1) 322 × 13: 966, 322, 4186
(2) 322 × 33: 966, 966, 10626
(3) 314 × 14: 1256, 314, 4396
(4) 314 × 45: 1570, 1256, 14130
(5) 407 × 24: 1628, 814, 9768
(6) 423 × 51: 423, 2115, 21573
(7) 906 × 37: 6342, 2718, 33522
(8) 316 × 70 = 22120
(9) 609 × 55: 3045, 3045, 33495
(10) 370 × 38: 2960, 1110, 14060
(11) 135 × 16: 810, 135, 2160
(12) 534 × 47: 3738, 2136, 25098
(13) 412 × 67: 2884, 2472, 27604
(14) 619 × 58: 4952, 3095, 35902
(15) 270 × 50 = 13500
(16) 164 × 56: 984, 820, 9184
(17) 672 × 32: 1344, 2016, 21504
(18) 608 × 89: 5472, 4864, 54112
(19) 731 × 44: 2924, 2924, 32164
(20) 345 × 95: 1725, 3105, 32775

① Complete the passage using vocabulary words defined below.

The sloth is an animal that lives up to its name, which means "laziness." These (1) mammals are (2) infamous for being slow and sleeping up to twenty hours a day. Sloths (3) dwell in the trees of the tropical forests in Central and South America. While their (4) lengthy arms and wooly fur make them look like monkeys, they are more closely related to armadillos and anteaters. There are two main (5) species of sloth. Sloths with two toes hang upside-down, while sloths with three toes like to sit upright. Three-toed sloths also have an extra (6) vertebrae in their necks so they can turn their heads almost all the way around. Both types of sloth are slow. In fact, they're so slow that (7) algae grows on their fur. Some scientists think that sloths move slow so (8) predators won't see them. The green algae also acts as (9) camouflage. But they're not only slow moving—a sloth can take up to a month to (10) digest one meal.

infamous	have a bad reputation
vertebrae	a section of bone or cartilage that make up the spinal column
dwell	to stay for a while; to live in a place
algae	any plant or plantlike living creature similar to seaweed
species	a category of living things; a class of things of the same kind and with the same name
digest	to break down food and absorb it in the body
predators	animals that lives by killing and eating other animals
camouflage	the hiding or disguising of something by covering it up or changing the way it looks
mammals	warm-blooded animals with verterbrae that feed their babies with window.
lengthy	very long

DAY 5, pages 9 & 10

1 Multiply.

(1) 1140 × 2 = 2280
(2) 1273 × 3 = 3819
(3) 2124 × 4 = 8496
(4) 2315 × 5 = 11575
(5) 3112 × 6 = 18672
(6) 1307 × 7 = 9149
(7) 1005 × 8 = 8040
(8) 1084 × 9 = 9756
(9) 2004 × 6 = 12024
(10) 2107 × 4 = 8428
(11) 132 × 123: 396, 264, 132, 16236
(12) 115 × 134: 460, 345, 115, 15410
(13) 122 × 146: 732, 488, 122, 17812
(14) 213 × 158: 1704, 1065, 213, 33654
(15) 201 × 113: 603, 201, 201, 22713
(16) 230 × 125: 1150, 460, 230, 28750
(17) 216 × 107: 1512, 2160, 23112
(18) 204 × 109: 1836, 2040, 22236
(19) 270 × 261: 270, 1620, 540, 70470
(20) 163 × 310: 1630, 489, 50530

① Complete the passage using vocabulary words defined below.

Get a ping-pong ball, a rubber ball, and a wooden ball of the same size. Place all three in water. Ever (1) ponder why the ping-pong ball floats the best and the wooden ball is almost underwater? It's easy to see that light objects filled with air, like (2) inflatable rafts, are good at floating. But solid things can float too. Why? A Greek (3) scientist named Archimedes, who was born around 287 BC, was able to explain this (4) phenomenon. Archimedes' first (5) theory was that floating objects are held up by a thrust called (6) buoyancy. His second idea was that the force needed to keep the object (7) afloat was equal to how much water the object (8) displaced. After a lot of (9) experiments, Archimedes could (10) prove that the amount of upward force on a floating object is equal to the weight of the water it moves out of place.

ponder	think about something carefully
inflatable	able to fill with air or gas
phenomenon	a fact, feature, or event of scientific interest
scientist	a person skilled in science
theory	an idea that is the starting point for argument or investigation
experiments	tests; operations carried out in order to discover something
afloat	carried on, or as if on, the water
buoyancy	the tendency to float or to rise when in a fluid
displaced	removed from an usual or proper place
prove	to show the truth by evidence

DAY 6, pages 11 & 12

1 Multiply.

(1) 61 × 8 = 488
(2) 572 × 7 = 4004
(3) 47 × 5 = 235
(4) 66 × 8 = 528
(5) 245 × 4 = 980
(6) 809 × 7 = 5663
(7) 3204 × 5 = 16020
(8) 6552 × 9 = 58968
(9) 104 × 9 = 936
(10) 753 × 7 = 5271
(11) 69 × 29: 621, 138, 2001
(12) 105 × 24: 420, 210, 2520
(13) 246 × 53: 738, 1230, 13038
(14) 93 × 39: 837, 279, 3627
(15) 741 × 56: 4446, 3705, 41496
(16) 230 × 80 = 18400
(17) 87 × 30 = 2610
(18) 618 × 73: 1854, 4326, 45114
(19) 50 × 46: 300, 200, 2300
(20) 208 × 319: 1872, 208, 624, 66352

① Complete the crossword puzzle using the sentences below as clues.

							(8)									(10)
		(3) c	o	l	o		g	n	e							t
					(7)		e						(9)			h
		(6)			r		n	(5) w	r	e	c	k	a	g	e	
		h			a		e					n				o
		i		(2)	p	a	r	a	c	h	u	t	e			r
		g			i		o						e			y
		h			d		u						l			
(1)	d	w	e	l	l	(4)	s	p	e	c	i	e	s			
		a			y		l									
		y					y									

ACROSS
(1) Groups of monkeys still ______ in these forests.
(2) The stunt woman opened her ______ about 800 meters or 2600 feet from the ground.
(3) The man had sprayed on too much ______.
(4) A new ______ of frog was discovered by scientists.
(5) The man was saved from the ______ of the crashed car.

DOWN
(6) We got on the wrong ______ and drove in the wrong direction.
(7) I ______ finished my test so I rechecked my answers.
(8) The chef ______ crushed pepper over the salad.
(9) Her family ______ in church when they are praying.
(10) Our music teacher tried to explain the ______ behind harmonies.

DAY 7, pages 13 & 14

1 Divide.

(1) 14 ÷ 2 = 7
(2) 25 ÷ 2 = 12 R 1
(3) 30 ÷ 2 = 15
(4) 54 ÷ 2 = 27
(5) 24 ÷ 3 = 8
(6) 28 ÷ 3 = 9 R 1
(7) 36 ÷ 3 = 12
(8) 32 ÷ 4 = 8
(9) 41 ÷ 4 = 10 R 1
(10) 56 ÷ 4 = 14
(11) 40 ÷ 5 = 8
(12) 43 ÷ 5 = 8 R 3
(13) 60 ÷ 5 = 12
(14) 30 ÷ 6 = 5
(15) 55 ÷ 6 = 9 R 1
(16) 78 ÷ 6 = 13
(17) 42 ÷ 7 = 6
(18) 57 ÷ 7 = 8 R 1
(19) 84 ÷ 7 = 12
(20) 48 ÷ 8 = 6
(21) 60 ÷ 8 = 7 R 4
(22) 90 ÷ 8 = 11 R 2
(23) 54 ÷ 9 = 6
(24) 65 ÷ 9 = 7 R 2
(25) 99 ÷ 9 = 11

① Read the short passage. Then choose words from the passage to complete the definitions below.

Canada is the second largest country in the world, but it only has half of one percent of the world's **population**. That means a lot of open space. Canada has lakes, rivers, mountains, **plains**, forests, and swamps. It even has the only **temperate** rain forest in the world. Canada **spans** more than half of the Northern **Hemisphere**. In the far north of Canada, you can see ice, snow, and **glaciers**.

With all this space comes many different animals—bears, mountain lions, otters, and many freshwater fish. Canadians **cherish** nature and wildlife. Forty-one national parks and three marine **conservation** areas have been made to protect animals like the wolf and **lynx**. These animals need to be protected because they have been **overhunted**.

(1) glaciers — large bodies of ice that move slowly
(2) cherish — to hold dear; to keep with care and affection
(3) conservation — a careful protection of something
(4) plains — broad areas of level or rolling treeless country
(5) lynx — a large wild cat
(6) population — the whole number of people living in a country or region
(7) overhunted — hunted too much
(8) temperate — a climate that is usually mild without very cold or hot temperatures
(9) spans — reaches or extends across
(10) hemisphere — half of the earth

DAY 8, pages 15 & 16

1 Divide.

(1) 224 ÷ 2 = 112
(2) 208 ÷ 2 = 104
(3) 150 ÷ 2 = 75
(4) 150 ÷ 3 = 50
(5) 315 ÷ 3 = 105
(6) 320 ÷ 3 = 106 R 2
(7) 140 ÷ 4 = 35
(8) 408 ÷ 4 = 102
(9) 350 ÷ 4 = 87 R 2
(10) 350 ÷ 5 = 70
(11) 570 ÷ 5 = 114
(12) 473 ÷ 5 = 94 R 3
(13) 630 ÷ 6 = 105
(14) 450 ÷ 6 = 75
(15) 726 ÷ 6 = 121
(16) 455 ÷ 7 = 65
(17) 721 ÷ 7 = 103
(18) 500 ÷ 7 = 71 R 3
(19) 840 ÷ 8 = 105
(20) 454 ÷ 8 = 56 R 6
(21) 616 ÷ 8 = 77
(22) 198 ÷ 9 = 22
(23) 354 ÷ 9 = 39 R 3
(24) 505 ÷ 9 = 56 R 1
(25) 963 ÷ 9 = 107

① Read the short passage. Then choose words from the passage to complete the definitions below.

A long time ago people learned something that would **alter** history: people learned how to **harness** fire. By striking stones together, a person could make a **spark**. Most likely, two **minerals** were used as **equipment** for starting fires. They gave off sparks when hit with something hard. The other **method** of creating fire was rubbing wooden sticks together. Just as your hands get warm when you rub them together, the **friction** of wood being rubbed together **generates** heat. **Tinder** would be put nearby to catch fire.

When a fire is lit, it creates light. The flame's color can tell you how hot the flame is and how much energy is being **released**. A bright blue flame is very hot and a dull yellow flame is cooler.

(1) method — a way, plan, or procedure for doing something
(2) spark — a bright flash; a small bit of burning material
(3) equipment — tools; necessary items used for a purpose
(4) minerals — natural materials usually from the ground
(5) generates — causes; brings into existence
(6) harness — to put to work; use
(7) tinder — a material that burns easily
(8) released — set free
(9) friction — the rubbing of one thing against another
(10) alter — change; to make different in some particular way

DAY 9, pages 17 & 18

1 Divide.

(1) 21) 46 = 2 R 4 (2) 21) 67 = 3 R 4 (3) 21) 105 = 5 (4) 31) 221 = 7 R 4 (5) 32) 235 = 7 R 11

(6) 43) 120 = 2 R 34 (7) 47) 300 = 6 R 18 (8) 56) 290 = 5 R 10 (9) 53) 350 = 6 R 32 (10) 64) 470 = 7 R 22

(11) 67) 600 = 8 R 64 (12) 71) 362 = 5 R 7 (13) 76) 532 = 7 (14) 82) 422 = 5 R 12 (15) 88) 528 = 6

(16) 91) 275 = 3 R 2 (17) 94) 658 = 7 (18) 74) 296 = 4 (19) 85) 455 = 5 R 30 (20) 96) 864 = 9

① Read the passage. Then choose words from the passage to answer the questions below.

A mouse happened to run into the mouth of a sleeping lion, who awoke with a jolt. He pulled the frightened mouse from his mouth and was just about put him back in, when the little fellow began begging the lion to let him go. The mouse said, "If you spare my life, I shall be grateful forever and pay you back some day." The lion replied, "Haha! What good could a tiny mouse do—except to whet my appetite." But the lion thought the idea was so funny that he let the mouse go for giving him a good laugh and because the mouse was nothing more than a pre-snack snack to him.

Later that same day, the lion was running through the plains when he was caught by some hunters and bound by ropes to a tree. The mouse, hearing his roars and groans, came quickly. By gnawing the ropes, he was able to set the lion free, saying, "You laughed at me once, as if you could receive no return from me, but now, you see, it is you who have to be grateful to me." When there is a turn of events, even the most powerful can owe something to the weak.

(1) Who catches the lion?
Some hunters catch the lion.

(2) Where is the lion running when he is caught?
The lion is running through the plains.

(3) What is the lion doing when the mouse runs into his mouth?
The lion is sleeping when the mouse runs into his mouth.

(4) When is the lion caught?
The lion is caught later that same day.

(5) Why does the mouse help the lion?
The mouse helps the lion because the lion spared the mouse's life.

(6) How does the mouse help the lion escape?
The mouse helps the lion escape by gnawing the ropes.

DAY 10, pages 19 & 20

1 Divide.

(1) 21) 275 = 13 R 2 (2) 21) 665 = 31 R 14 (3) 27) 631 = 23 R 10 (4) 31) 335 = 10 R 25 (5) 31) 639 = 20 R 19

(6) 36) 989 = 27 R 17 (7) 43) 500 = 11 R 27 (8) 44) 903 = 20 R 23 (9) 47) 982 = 20 R 42 (10) 58) 815 = 14 R 3

(11) 21) 1357 = 64 R 13 (12) 41) 2222 = 54 R 8 (13) 53) 3571 = 67 R 20 (14) 65) 4456 = 68 R 36 (15) 71) 5248 = 73 R 65

(16) 83) 2713 = 32 R 57 (17) 97) 7324 = 75 R 49 (18) 23) 4996 = 217 R 5 (19) 33) 7133 = 216 R 5 (20) 42) 8534 = 203 R 8

① Read the passage. Then choose words from the passage to answer the questions below.

Once upon a time there was a prince who wanted to marry a princess, but she had to be a special princess. So he traveled east, then west, then north, and then south. There were plenty of princesses, but he could not find one that he considered special. In each case there was some little defect, which made him unsure. So he came home again in very low spirits. He thought he would be alone forever.

The night after the prince's return there was a dreadful storm; there was thunder and lightning and the rain streamed down in torrents. It was fearsome! There was a knocking heard at the palace gate, and the old king and queen went to open it.

There stood a princess outside the gate; but oh, she looked dreadful from the rain and the storm! The water was running down from her hair and her dress into the points of her shoes and out at the heels again. Her hair was a mess, whipping this way and that way from the wind. But she said she was a princess and had come to marry the prince.

(1) Who answered the door of the palace?
The king and queen answered the door of the palace.

(2) What was the prince searching for?
The prince was searching for a princess.

(3) Where did the prince travel?
The prince traveled east, then west, then north, and then south.

(4) When was the storm?
The storm was the night after the prince returned.

(5) Why was the prince in low spirits?
The prince was in low spirits because he thought he would be alone forever.

(6) How did the princess look outside the gate?
The princess looked dreadful.

DAY 11, pages 21 & 22

1 Divide.

(1) 121) 537 = 4 R 53 (2) 173) 888 = 5 R 23 (3) 247) 935 = 3 R 194 (4) 307) 935 = 3 R 14

(5) 324) 826 = 2 R 178 (6) 257) 8784 = 34 R 46 (7) 391) 8901 = 22 R 299 (8) 754) 9004 = 11 R 710

(9) 832) 9872 = 11 R 720 (10) 903) 9879 = 10 R 849

① Read the passage. Then choose words from the passage to answer the questions below.

"Well, we shall soon find out!" thought the queen. But she said nothing and snuck into the guest bedroom. She took off all the bed linens and laid a pea on the mattress. Then she put twenty more mattresses on top of the pea and twenty quilts on the top of the mattresses. And this was the bed on which the princess was to sleep.

The next morning the queen asked how the princess had slept.

"Oh, very badly!" said the princess. "I scarcely closed my eyes all night! I don't mean to be ungrateful, but I don't know what was in the bed. I laid on something so hard that my whole body is black and blue. It is worse than the storm I escaped!"

Now they saw that she was a truly special princess because she had felt the pea through the twenty mattresses and the twenty quilts. Only a true princess could be so sensitive.

So the prince asked to marry her that very moment, and the pea was put into the Royal Museum, where it still can be seen today.

(1) Who laid the pea on the bed?
The queen laid the pea on the bed.

(2) What was on top of the pea?
Twenty more mattresses and twenty quilts were on top of the pea.

(3) Where did the princess sleep?
The princess slept in the guest bedroom.

(4) When did the prince ask the princess to marry him?
The prince asked the princess to marry him that very moment.

(5) Why did the queen put the pea on the bed?
The queen put a pea on the bed to find out if the girl was a true princess.

(6) How did the prince know that she was a truly special princess?
The prince knew because only a true princess could be so sensitive.

DAY 12, pages 23 & 24

1 Divide.

(1) 55) 375 = 6 R 45 (2) 7) 60 = 8 R 4 (3) 8) 5846 = 730 R 6 (4) 38) 621 = 16 R 13 (5) 52) 8061 = 155 R 1

(6) 5) 805 = 161 (7) 18) 550 = 30 R 10 (8) 63) 9036 = 143 R 27 (9) 9) 7001 = 777 R 8 (10) 42) 720 = 17 R 6

(11) 29) 1506 = 51 R 27 (12) 123) 222 = 1 R 99 (13) 251) 628 = 2 R 126 (14) 165) 5348 = 32 R 68 (15) 18) 5248 = 291 R 10

(16) 328) 1810 = 5 R 170 (17) 97) 732 = 7 R 53 (18) 257) 8993 = 34 R 255 (19) 482) 7008 = 14 R 260 (20) 142) 9500 = 66 R 128

① Read the passage. Then choose words from the passage to answer the questions below.

Vitamins are needed by animals and plants for nutrition, growth, and life. Different vitamins have different jobs. Henrik Dam and Edward Doisy discovered vitamin K in 1934. Vitamin K is a vitamin that helps blood thicken and set. When blood sets, it forms a clot or lump and stops any bleeding.

Dam and his team discovered vitamin K by studying chicks that weren't well fed and bled easily. If the chicks had a cut it would also take a long time for the bleeding to stop. Dam believed that the chicks were missing a vitamin in their food that helped their blood clot. He found out this vitamin comes from green leaves and named it vitamin K. Dam and Doisy were able to find the vitamin in an alfalfa plant which has green leaves in a clover shape and blooms a blueish flower. They could make the chicks' blood clot better and faster by feeding them vitamin K.

In 1943, both scientists were awarded the Nobel Prize for Medicine for their research. This famous award is given each year in Stockholm, Sweden.

(1) Who discovered vitamin K?
Henrik Dam and Edward Doisy discovered vitamin K.

(2) What does vitamin K do?
Vitamin K helps blood thicken and set.

(3) Where is the Nobel Prize given each year?
The Nobel Prize is given in Stockholm, Sweden.

(4) When did Dam and Doisy discover vitamin K?
Dam and Doisy discovered vitamin K in 1934.

(5) Why did the chicks bleed easily?
The chicks bled easily because they were missing vitamin K.

(6) How could the scientists make the chicks' blood clot better and faster?
The scientists could make the chicks blood clot by feeding them vitamin K.

DAY 13, pages 25 & 26

0 0.1 0.2 0.3 0.4 0.5 0.6 0.7 0.8 0.9 1 1.1 1.2

1 Write the correct decimal in each box.

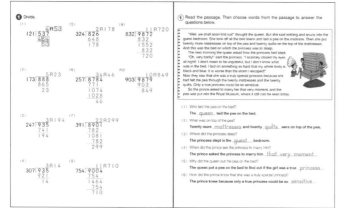

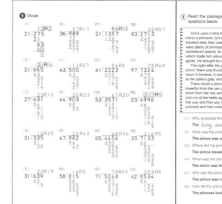

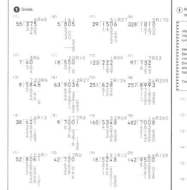

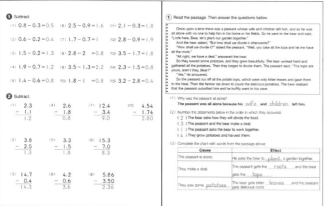

(1) 0.2, 0.4, 0.6, 0.8
(2) 0.1, 0.3, 0.5, 0.7, 0.9
(3) 0.2, 0.6, 0.9, 1.1
(4) 0.1, 0.5, 0.8
(5) 0.9, 1.4, 1.7, 2.2
(6) 0.7, 1.1, 1.5, 2.3

① Read the passage. Then read the sentences below. Circle the "T" if the sentence is true. Circle the "F" if the sentence is false.

After World War II, many nations decided to form a group to assure people's freedom and to work towards peace around the world. On October 24, 1945, this group was formed and it was called the United Nations. Now that day is celebrated around the world as United Nations Day. The United Nations had many goals, but they started with four main aims: to set up and maintain international peace, to grow friendships between countries, to help countries work together to fix problems, and to convince nations to respect human rights and freedoms. Fifty-one countries joined together initially to create the United Nations. As of 2011, 193 nations are members.

The United Nations has its own peacekeeping force, which includes members of the military, police, and general public who work to build peace in countries that have conflicts. Peacekeepers have been sent to countries all over the world to protect citizens and restore peace.

(1) Before World War II, many nations formed the United Nations. — F
(2) The United Nations was formed on October 24, 1945. — T
(3) October 24th is United Nations Day. — T
(4) The United Nations started with eight main goals. — F
(5) At first, fifty-one countries made up the United Nations. — T
(6) 190 nations were members as of 2011. — F
(7) One of the United Nations' goals is to grow friendships between countries. — T
(8) The peacekeepers are members of the military, police, judges and general public. — F
(9) The peacekeeping force builds peace in countries that have conflicts. — T
(10) Peacekeepers have been sent all over the world. — T

DAY 14, pages 27 & 28

1 Add.

(1) 1 + 0.5 = 1.5 (6) 0.2 + 0.7 = 0.9 (11) 1.3 + 2 = 3.3
(2) 2 + 0.7 = 2.7 (7) 0.4 + 0.6 = 1 (12) 1.9 + 1 = 2.9
(3) 1 + 1.5 = 2.5 (8) 0.6 + 1.7 = 2.3 (13) 1.4 + 1.2 = 2.6
(4) 1 + 0.9 = 1.9 (9) 1.3 + 0.5 = 1.8 (14) 2.1 + 1.5 = 3.6
(5) 2 + 1.3 = 3.3 (10) 1.5 + 0.7 = 2.2 (15) 1.8 + 1.7 = 3.5

2 Add.

(1) 1.3 + 2.5 = 3.8 (2) 2.3 + 0.4 = 2.7 (3) 14.5 + 1.5 = 16
(4) 3.6 + 12.1 = 15.7 (5) 0.8 + 11.7 = 12.5 (6) 8.0 + 12.5 = 20.5
(7) 2.05 + 1.60 = 3.65 (8) 0.68 + 2.40 = 3.08 (9) 6.24 + 1.75 = 7.99
(10) 5.52 + 14.48 = 20

① Read the passage. Then read the sentences below. Circle the "T" if the sentence is true. Circle the "F" if the sentence is false.

Remote-control spacecrafts have been flying around space for more than forty years. These explorers are "unmanned," meaning there is no person onboard the spacecraft. These crafts travel around like a stone from a slingshot. These amazing robots have gone as far as Mercury, Venus, Mars, Jupiter, Saturn, Uranus, and Neptune to get data and pictures. They are our eyes and ears in places where people cannot go. The spacecraft uses each planet's gravity to pull them in and shoot them onward. Gravity is the force that holds objects down on the surface of the earth.

Yet none of these journeys would be possible without the Deep Space Network, which is a system of antennas. Antennas are devices that send and receive signals. These signals can travel up to billions of miles or kilometers. The farther a spacecraft has to go, the larger the antenna needs to be. Some antennas can be as large as a two-story house or even larger!

(1) Remote-control spacecrafts have been flying around space for more than four decades. — T
(2) Spacecrafts use a planet's gravity like a slingshot to pull them in and shoot them onward. — T
(3) Remote-control spacecrafts have gone as far as Neptune. — T
(4) Without the Deep Space Network, we would not be able to have so many television channels. — F
(5) Unmanned spacecrafts take pictures and get data. — T
(6) The Deep Space Network is a system of antennas. — T
(7) Remote-control spacecrafts always have at least one pilot on board. — F
(8) Antennas are devices that only receive signals. — F
(9) The farther a spacecraft has to go, the larger the antenna needs to be. — T
(10) Signals can travel only one million miles. — F

DAY 15, pages 29 & 30

1 Subtract.

(1) 0.8 − 0.3 = 0.5 (6) 2.5 − 0.9 = 1.6 (11) 2.1 − 0.3 = 1.8
(2) 0.6 − 0.2 = 0.4 (7) 1.7 − 0.7 = 1 (12) 2.8 − 0.9 = 1.9
(3) 1.5 − 0.2 = 1.3 (8) 2.8 − 2 = 0.8 (13) 3.5 − 1.7 = 1.8
(4) 1.9 − 0.7 = 1.2 (9) 3.5 − 1.3 = 2.2 (14) 2.3 − 1.5 = 0.8
(5) 1.4 − 0.6 = 0.8 (10) 1.8 − 1 = 0.8 (15) 3.2 − 2.8 = 0.4

2 Subtract.

(1) 2.3 − 1.1 = 1.2 (2) 3.8 − 2.5 = 1.3 (3) 14.7 − 0.4 = 14.3
(4) 2.6 − 1.8 = 0.8 (5) 3.3 − 1.5 = 1.8 (6) 4.2 − 0.6 = 3.6
(7) 12.4 − 3.4 = 9.0 (8) 15.3 − 7.0 = 8.3 (9) 5.86 − 3.50 = 2.36
(10) 4.54 − 1.74 = 2.80

① Read the passage. Then answer the questions below.

Once upon a time there was a peasant whose wife and children left him, and so he was all alone with no one to help him in his home or his fields. So he went to the bear and said, "Look here, Bear, let's plant our garden together."

And the bear asked, "But how shall we divide it afterwards?"

"How shall we divide it?" asked the peasant. "Well, you take all the tops and let me have all the roots."

"All right, we have a deal," answered the bear.

So they sowed some potatoes, and they grew beautifully. The bear worked hard and gathered all the potatoes. Then they began to divide them. The peasant said, "The tops are yours, aren't they, Bear?"

"Yes," he answered.

So the peasant cut off all the potato tops, which were only bitter leaves and gave them to the bear. Then the farmer sat down to count the delicious potatoes. The bear realized that the peasant outwitted him and he huffily went to his cave.

(1) Why was the peasant all alone?
The peasant was all alone because his wife and children left him.

(2) Number the statements below in the order in which they occurred.
(2) The bear asks how they will divide the food.
(3) The peasant and the bear make a deal.
(1) The peasant asks the bear to work together.
(4) They grow potatoes and harvest them.

(3) Complete the chart with words from the passage above.

Cause	Effect
The peasant is alone.	He asks the bear to plant a garden together.
They make a deal.	The peasant gets the roots, and the bear gets the tops.
They sow some potatoes.	The bear gets bitter leaves and the peasant gets delicious roots.

DAY 16, pages 31 & 32

1 Rewrite the improper fractions as mixed numbers or whole numbers.

(1) $\frac{6}{5} = 1\frac{1}{5}$ (6) $\frac{5}{4} = 1\frac{1}{4}$ (11) $\frac{16}{7} = 2\frac{2}{7}$
(2) $\frac{8}{5} = 1\frac{3}{5}$ (7) $\frac{9}{4} = 2\frac{1}{4}$ (12) $\frac{13}{8} = 1\frac{5}{8}$
(3) $\frac{13}{5} = 2\frac{3}{5}$ (8) $\frac{7}{7} = 1$ (13) $\frac{19}{8} = 2\frac{3}{8}$
(4) $\frac{6}{6} = 1$ (9) $\frac{10}{7} = 1\frac{3}{7}$ (14) $\frac{10}{9} = 1\frac{1}{9}$
(5) $\frac{11}{6} = 1\frac{5}{6}$ (10) $\frac{13}{7} = 1\frac{6}{7}$ (15) $\frac{16}{9} = 1\frac{7}{9}$

2 Rewrite the mixed numbers and whole numbers as improper fractions.

(1) $1 = \frac{4}{4}$ (5) $2\frac{1}{5} = \frac{11}{5}$ (9) $1\frac{3}{7} = \frac{10}{7}$
(2) $1\frac{2}{5} = \frac{7}{5}$ (6) $1\frac{1}{3} = \frac{4}{3}$ (10) $1\frac{3}{9} = \frac{12}{9}$
(3) $1\frac{1}{4} = \frac{5}{4}$ (7) $2\frac{2}{3} = \frac{8}{3}$
(4) $2 = \frac{10}{5}$ (8) $2 = \frac{12}{6}$

① Read the passage. Then answer the questions below.

The next spring the peasant again came to see the bear and said, "Look here, Bear, let's work together again, shall we?"

The bear remembered the potato disaster from the year before and answered, "Right-ho! Only this time, I'll make the deal: you can have the tops, and I'm going to have the roots!"

"Very well," said the peasant.

But this year they sowed some wheat, and when the ears grew up and ripened, you never saw such a sight. The bear worked hard and gathered all the wheat, and then they began to divide it. This time, the peasant took all the tops with the grain for baking bread and gave the bear the straw and the roots, which weren't much good for anything. The bear realized that the peasant had outwitted him again!

"Well, good-bye!" said the bear to the peasant. "I'm not going to work with you anymore. You're too crafty!" And with that he went off into the forest.

(1) Why did the bear make a different deal?
The bear made a different deal because he remembered the potato disaster from the year before.

(2) Number the statements below in the order in which they occurred.
(4) The bear gets the straw and roots.
(3) They grow wheat together.
(2) The bear makes a new deal.
(1) The peasant asks the bear to work together again.

(3) Complete the chart with words from the passage above.

Cause	Effect
The bear remembers last year's potato disaster.	He makes a new deal.
They make a deal.	The peasant gets the tops, and the bear gets the roots.
They plant wheat.	The bear gets the straw and roots, and the peasant gets the grain.

DAY 17, pages 33 & 34

1 Add.

(1) $\frac{2}{5}+\frac{1}{5}=\frac{\boxed{3}}{5}$	(11) $\frac{1}{5}+\frac{4}{5}=\frac{\boxed{5}}{5}=\boxed{1}$
(2) $\frac{2}{5}+\frac{2}{5}=\frac{4}{5}$	(12) $\frac{2}{7}+\frac{4}{7}=\frac{6}{7}$
(3) $\frac{3}{5}+\frac{1}{5}=\frac{4}{5}$	(13) $\frac{3}{7}+\frac{4}{7}=\frac{7}{7}=1$
(4) $\frac{2}{7}+\frac{1}{7}=\frac{3}{7}$	(14) $\frac{1}{9}+\frac{7}{9}=\frac{8}{9}$
(5) $\frac{3}{7}+\frac{1}{7}=\frac{4}{7}$	(15) $\frac{4}{9}+\frac{5}{9}=\frac{9}{9}=1$
(6) $\frac{1}{7}+\frac{4}{7}=\frac{5}{7}$	(16) $\frac{2}{11}+\frac{6}{11}=\frac{8}{11}$
(7) $\frac{2}{7}+\frac{3}{7}=\frac{5}{7}$	(17) $\frac{3}{11}+\frac{5}{11}=\frac{8}{11}$
(8) $\frac{2}{9}+\frac{2}{9}=\frac{4}{9}$	(18) $\frac{4}{11}+\frac{7}{11}=\frac{11}{11}=1$
(9) $\frac{1}{9}+\frac{4}{9}=\frac{5}{9}$	(19) $\frac{5}{11}+\frac{4}{11}=\frac{9}{11}$
(10) $\frac{4}{9}+\frac{3}{9}=\frac{7}{9}$	(20) $\frac{2}{7}+\frac{5}{7}=\frac{7}{7}=1$

1 Read the passage. Then answer the questions below.

"Wetland" is the name for any area of land that is between dry land and water, like swamps and bogs. When you think of wetlands you may think of mud, annoying mosquitoes, and stinky odors. People have destroyed many wetlands because they didn't know their value. More than half of the wetlands in the United States have been drained, filled, or used for the disposal of garbage. However, wetlands are a very important natural resource. Wetlands are similar to rain forests and coral reefs because they are home to many different animals, plants, and fish. Certain animals, like the wood stork, are endangered because of wetland destruction. Without wetlands, these animals will no longer be able to live. Wetlands also act like a sponge and soak up flooding water, rain, or melting snow, thereby protecting people and land.

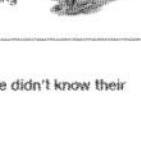

Because wetlands are in danger, animals, plants, land, and people are in danger, too. Therefore, environmental groups are starting programs to save the wetlands. Some governments are also passing laws to protect these areas.

(1) Why have people destroyed many wetland areas?
People have destroyed many wetland areas because people didn't know their value.

(2) How are wetlands similar to rain forests and coral reefs?
Wetlands are similar to rain forests and coral reefs because they are home **to many different** animals, plants, **and** fish.

(3) Why is the wood stork endangered?
The wood stork is endangered because of wetland destruction.

(4) Complete the chart with words from the passage above.

Cause	Effect
Wetlands soak up water.	Wetlands protect people and land from flooding.
The destruction of wetlands	Certain animals will no longer be able to live.
Because wetlands are in danger	Environmental groups are starting programs to save the wetlands.

DAY 18, pages 35 & 36

1 Subtract.

(1) $\frac{2}{5}-\frac{1}{5}=\frac{\boxed{1}}{5}$	(11) $1-\frac{3}{5}=\frac{\boxed{2}}{5}$
(2) $\frac{4}{5}-\frac{2}{5}=\frac{2}{5}$	(12) $\frac{7}{5}-\frac{4}{5}=\frac{3}{5}$
(3) $\frac{4}{5}-\frac{1}{5}=\frac{3}{5}$	(13) $\frac{8}{7}-\frac{5}{7}=\frac{3}{7}$
(4) $\frac{2}{7}-\frac{1}{7}=\frac{1}{7}$	(14) $1-\frac{3}{7}=\frac{4}{7}$
(5) $\frac{3}{7}-\frac{1}{7}=\frac{2}{7}$	(15) $\frac{8}{9}-\frac{5}{9}=\frac{3}{9}$
(6) $\frac{4}{7}-\frac{2}{7}=\frac{2}{7}$	(16) $\frac{10}{11}-\frac{6}{11}=\frac{4}{11}$
(7) $\frac{6}{7}-\frac{2}{7}=\frac{4}{7}$	(17) $1-\frac{5}{8}=\frac{3}{8}$
(8) $\frac{4}{9}-\frac{2}{9}=\frac{2}{9}$	(18) $\frac{13}{11}-\frac{7}{11}=\frac{6}{11}$
(9) $\frac{5}{9}-\frac{1}{9}=\frac{4}{9}$	(19) $1-\frac{4}{9}=\frac{5}{9}$
(10) $\frac{7}{9}-\frac{5}{9}=\frac{2}{9}$	(20) $1-\frac{2}{11}=\frac{9}{11}$

1 Read the passage. Then answer the questions below.

Most of our electricity comes from burning coal, oil, or gas, but there is a limited amount of these fuels, and one day there will be none left. Therefore, many people are looking for new ways to create electricity from resources that won't run out—in other words, a source of renewable energy.

People also want to find new types of energy because burning coal, oil, and gas pollutes the earth. Scientists are studying ways to make energy out of sunlight, wind power, and water.

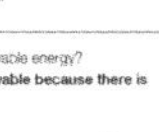

The most common renewable source of electricity is hydropower. Hydropower is popular because it is not very expensive to produce. Hydropower can be created by using the water from a waterfall. Another way is by using a river and a special dam (a barrier that controls the flow of water in a river). When water from a river passes through a dam, the water turns a machine that looks like a fan. This movement creates energy that can be captured and turned into electricity. By controlling the flow of water, people can produce more or less electricity.

(1) Why is electricity made from burning coal not considered renewable energy?
Electricity made from burning coal is not considered renewable because there is a limited **amount of coal.**

(2) Why are people looking for new types of energy?
People are looking for new types of energy because coal, oil, and gas pollutes **the earth.**

(3) Why is hydropower popular?
Hydropower is popular because it is not very expensive **to produce.**

(4) Complete the chart with words from the passage above.

Cause Effect	Effect
Limited amounts of coal, oil, and gas	People are looking for new ways to create electricity.
Water from a river goes through a hydropower dam	The water turns a machine that looks like a fan
Controling the flow of water	People can produce more or less electricity.

DAY 19, pages 37 & 38

1 Read the word problem and write the number sentence below. Then answer the question.

(1) A box of pencils includes 12 pencils. If John bought 7 boxes, how many pencils did he buy?

Number of pencils per box / Number of boxes / Total pencils
12 × 7 = 84

Ans. 84 pencils

(2) Robin's school has 25 classes. There are 30 students in each class. How many students are in Robin's school?

25 × 30 = 750

Ans. 750 students

(3) Colored pencils are sold in packs of 8. Your class has 28 packs. How many colored pencils does your class have?

28 × 8 = 224

Ans. 224 colored pencils

(4) The gardener gave each student 15 seeds to plant. If there are 27 students, how many seeds did the gardener give away?

15 × 27 = 405

Ans. 405 seeds

(5) Stickers are sold in rolls of 24. If Julie buys 11 rolls, how many stickers does she have?

24 × 11 = 264

Ans. 264 stickers

1 Read the passage. Then answer the questions below.

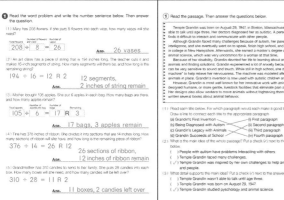

Halle Berry is an admired public figure in American history. She was the first African-American woman to win an Academy Award for Best Actress. The Academy Awards are presented each year by the Academy of Motion Picture Arts and Sciences to recognize achievements in film.

Halle Berry was born on August 14, 1966, in Cleveland, Ohio. She was a teenage finalist in national beauty pageants and went on to work as a model. She began acting when she was twenty-three years old. She was cast in many different roles and earned a lot of praise for her work.

In 1999, Berry even starred in a film about the historic movie star Dorothy Dandridge, who was the first African American to be nominated for the Academy Award for Best Actress. When Berry won the Academy Award in 2001, she said, "This moment is so much bigger than me. This moment is for Dorothy Dandridge, Lena Horne, [and] Diahann Carroll," and she went on to dedicate the award to actresses of color.

(1) Read each title below. For which paragraph would each make a good title? Draw a line to connect each to the appropriate paragraph.
(a) Halle Berry's Childhood — (ii) Second paragraph
(b) Halle Berry's Tribute to African-American Actresses — (iii) Third paragraph
(c) Introduction to Halle Berry and the Academy Awards — (i) First paragraph

(2) Put a check (✓) next to the best title for the whole passage below.
( ) How to Break Into Acting
(✓) The Biography of Halle Berry
( ) The Academy Awards

DAY 20, pages 39 & 40

1 Read the word problem and write the number sentence below. Then answer the question.

(1) We had 350 inches of ribbon for our group. We divided the ribbon equally among 7 students. How much ribbon did each student get?

Total ribbon / Number of students / Ribbon per student
350 ÷ 7 = 50

Ans. 50 inches of ribbon

(2) There are 204 roses. The florist split them evenly into 12 bunches. How many roses are in each bunch?

204 ÷ 12 = 17

Ans. 17 roses

(3) The cafeteria has 265 apples. They are divided into 8 boxes equally. How many apples are in each box, and how many apples remain?

Total apples / Number of boxes / Apples per box / Remaining
265 ÷ 8 = 33 R 1

Ans. 33 apples, 1 apple remains

(4) Robin has 186 lollipops at his party. He divides them equally among 21 people. How many lollipops does each person get, and how many lollipops remain?

186 ÷ 21 = 8 R 18

Ans. 8 lollipops, 18 lollipops remain

(5) The art teacher has 113 sheets of colored paper. She divides them equally among 24 students. How many sheets of colored paper does each student get?

113 ÷ 24 = 4 R 17

Ans. 4 sheets of colored paper

1 Read the passage. Then answer the questions below.

A long time ago, Greeks believed that all liquids were made up of mostly water. However, scientists have discovered that all liquids are made up of particles called atoms. The smallest unit of water is a cluster of only three atoms.

Liquids can adapt to all different kinds of situations. Liquid can be thinly spread out, like when it spills across a table; or it can be tightly packed together, like when it is held in a bottle. When liquid is heated, the spaces between the particles expand and, so does the liquid. The opposite also occurs when a liquid is cooled—the particles contract and get closer together.

The tiny particles that make up a liquid are also attracted to each other and tend to keep close together. This attraction creates tension between the particles, which is why when you fill up a cup to the top, the water at the surface holds tight like the skin of a balloon. This tension also allows very light insects to walk on water. This phenomenon is called surface tension.

Liquids are also very powerful. Given enough time, liquids can wear away solid surfaces, like rocks. For example, a canyon is a deep and steep valley that has been carved out by a river. These kinds of valleys are often located where the river has a strong current that runs rapidly.

(1) Read each title below. For which paragraph would each make a good title? Draw a line to connect each title to the appropriate paragraph.
(a) The Discovery of Atoms — (i) First paragraph
(b) Surface Tension — (iii) Third paragraph
(c) The Power of Liquids — (iv) Fourth paragraph
(d) Liquid in Different Forms — (ii) Second paragraph

(2) Put a check (✓) next to the sentence that describes the main message of the fourth paragraph.
( ) Liquids can adapt. (✓) Liquids are powerful
( ) Canyons are made by liquid. ( ) When liquid cools, the particles contrast.

(3) What is the main idea of the whole passage? Put a check (✓) next to the correct idea below.
( ) Liquids are made up of atoms. ( ) Liquids are unchanging.
(✓) Liquids have adaptable qualities. ( ) The Greeks didn't know about atoms.

DAY 21, pages 41 & 42

1 Read the word problem and write the number sentence below. Then answer the question.

(1) Mary has 208 flowers. If she puts 8 flowers into each vase, how many vases will she need?

Total flowers / Number of flowers per vase / Number of vases
208 ÷ 8 = 26

Ans. 26 vases

(2) An art class has a piece of string that is 194 inches long. The teacher cuts it and makes 16-inch segments of string. How many segments will there be, and how long is the remaining string?

194 ÷ 16 = 12 R 2

Ans. 12 segments, 2 inches of string remain

(3) Mother bought 105 apples. She put 6 apples in each bag. How many bags are there, and how many apples remain?

Total apples / Number of apples per bag / Number of bags / Remaining
105 ÷ 6 = 17 R 3

Ans. 17 bags, 3 apples remain

(4) Tina has 376 inches of ribbon. She divides it into sections that are 14 inches long. How many sections of ribbon will she have, and how long is the remaining piece of ribbon?

376 ÷ 14 = 26 R 12

Ans. 26 sections of ribbon, 12 inches of ribbon remain

(5) Grandmother has 310 candies to send to her family. She puts 28 candies into each box. How many boxes will she need, and how many candies will be left over?

310 ÷ 28 = 11 R 2

Ans. 11 boxes, 2 candies left over

1 Read the passage. Then answer the questions below.

Temple Grandin was born on August 29, 1947, in Boston, Massachusetts. Grandin wasn't able to talk until age three. Her doctors diagnosed her as autistic. A person with autism often finds it difficult to interact and communicate with other people.

Although Grandin faced many challenges because of autism, her parents nurtured her intelligence, and she eventually went on to speak, finish high school, and study psychology in college in New Hampshire. Afterwards, she earned a master's degree and a doctorate in animal science, which was very uncommon for a woman at that time.

Because of her disability, Grandin devoted her life to learning about anxiety in people and animals and finding solutions. Grandin experienced a lot of anxiety because autistic people can be very sensitive to sound and touch. While still in high school, she designed a "squeeze machine" to help relieve her nervousness. The machine was modeled after a chute that held animals in place. Grandin's invention is now used with autistic children and adults.

However, Grandin is most well known for her innovative work with animals. She has designed humane, or more gentle, livestock facilities that eliminate pain and fear in animals. Her designs also allow workers to move animals without frightening them. She has also written several books about animal behavior.

(1) Read each title below. For which paragraph would each make a good title? Draw a line to connect each title to the appropriate paragraph.
(a) Grandin's First Invention — (iii) Third paragraph
(b) Being Diagnosed with Autism — (i) First paragraph
(c) Grandin's Legacy with Animals — (iv) Fourth paragraph
(d) Grandin Succeeds at School — (ii) Second paragraph

(2) What is the main idea of the whole passage? Put a check (✓) next to the correct idea below.
( ) People with autism have problems interacting with others.
( ) Temple Grandin faced many challenges.
(✓) Temple Grandin was inspired by her own challenges to help anxious animals and people.

(3) What detail supports the main idea? Put a check (✓) next to the answer.
( ) Temple Grandin wasn't able to talk until age three.
( ) Temple Grandin was born on August 29, 1947.
(✓) Temple Grandin studied pyschology and animal science.

DAY 22, pages 43 & 44

1 Read the word problem and write the number sentence below. Then answer the question.

(1) Kate has 54 dimes for her collection. Her younger sister has 18 dimes. How many times more dimes does Kate have than her sister?

Number of Kate's dimes / Number of Kate's sister's dimes / How many times more
54 ÷ 18 = 3

Ans. 3 times more dimes

(2) The grocer has 112 oranges in a box and 14 oranges out front. How many times more oranges does he have in the box than he has out front?

112 ÷ 14 = 8

Ans. 8 times more oranges

(3) A red ribbon is 36 yards long. It is 3 times longer than the blue ribbon. How long is the blue ribbon?

Length of red ribbon / How many times longer / Length of blue ribbon
36 ÷ 3 = 12

Ans. 12 yards long

(4) The bear at the zoo weighs 560 pounds. He weighs 7 times as much as the chimpanzee. How much does the chimpanzee weigh?

560 ÷ 7 = 80

Ans. 80 pounds

(5) The bedroom in Smith's house is 192 square feet, and the closet is 32 square feet. How many times bigger is the bedroom than the closet?

192 ÷ 32 = 6

Ans. 6 times bigger than the closet

1 Read the passage. Then answer the questions below.

Did you know that glass can be shattered with only the force of the human voice? This is because of the power of vibrations, or movement back and forth. The number of vibrations that an object makes each second is called its natural frequency. Anything that can vibrate—everything from a bridge to a violin string—has its own natural frequency. Just like a swing in a playground, if an object is given a push, it will move back and forth at its natural frequency and then gradually stop. But if you continue to push a swing in the right rhythm, it can rise higher and higher. This happens when you push according to the swing's natural frequency.

The same thing happens with the shattering glass. When someone sings, his or her voice creates a sound wave that vibrates. Different notes make different sound waves and thus vibrate at different rates. If a note is sung with a rate of vibration that matches the natural frequency of the glass, the glass could shatter. When the vibrations match, the energy from the voice transfers to the glass, and the powerful vibrations destroy the glass. This transfer is called "resonance."

Luckily, there is a way to demonstrate resonance without destroying glasses. By taking a wine glass and running a wet finger quickly around the rim of the glass, a person can create a note. If the person sings the same note aloud, the glass will resonate the note and the sound will become slightly louder.

(1) Read each title below. For which paragraph would it make a good title? Draw a line to connect each title to the appropriate paragraph.
(a) How a Note Can Destroy a Glass — (ii) Second paragraph
(b) An Experiment to Show Resonance — (iii) Third paragraph
(c) Introduction to Vibration — (i) First paragraph

(2) What is the main idea of the whole passage? Put a check (✓) next to the correct idea below.
( ) Science experiments are fun and informative.
(✓) Vibration and resonance can be a powerful force together.
( ) Everything has its own natural rate of vibration.

(3) What detail supports the main idea? Put a check (✓) next to the answer.
(✓) If you continue to push a swing at its natural frequency, it will rise.
( ) Many things—from bridges to violin strings—vibrate
( ) Singing loudly is not good for glasses.

DAY 23, pages 45 & 46

1 Read the word problem and write the number sentence below. Then answer the question.

(1) You used 0.2 pounds of sugar in your cake, and 1.7 pounds of sugar are left over. How much sugar was there in the beginning?

0.2 + 1.7 = 1.9

Ans. 1.9 pounds of sugar

(2) Julian's bag weighs 2.8 pounds. His father's bag is 1.2 pounds heavier than his. How much does his father's bag weigh?

2.8 + 1.2 = 4

Ans. 4 pounds

(3) Dan and Wendy were trying to throw a big rock. Dan threw it 1.3 meters. Wendy threw it 70 centimeters farther. How far did Wendy throw the rock?

70 cm = 0.7 m
1.3 + 0.7 = 2

Ans. 2 meters

(4) Ava's bag weighs 2.4 kilograms. Her sister's bag is 600 grams heavier. How much does her sister's bag weigh?

600 g = 0.6 kg
2.4 + 0.6 = 3

Ans. 3 kilograms

(5) Kelly had 2.1 liters of water in her water bottle. Her big water bottle can hold 800 milliliters more. How much can her big water bottle hold?

800 mL = 0.8 L
2.1 + 0.8 = 2.9

Ans. 2.9 liters

1 Read the passage. Then answer the questions below.

Roberto Clemente was one of the first Latin American baseball stars. He was born in a modest house in Puerto Rico on August 18, 1934. Clemente went on to become a twelve-time All Star and do important charity work in his free time.

At only fourteen years old, Clemente began playing softball on a men's team. By eighteen, he turned professional. In February of 1954, the Brooklyn Dodgers recruited Clemente but placed him in the minor leagues, where he didn't play very often. The Dodgers tried to hide his talent so other teams wouldn't want him. But it was too late—the Pittsburgh Pirates brought Clemente up to the major leagues.

Over eighteen seasons, Clemente collected impressive statistics and delighted baseball fans. No matter what kind of pitch, he could hit the ball. He had lightning speed, which made him a great base runner. He was also well known for his powerful and accurate throwing arm. Even towards the end of his career, Clemente continued to set records.

But one of the biggest challenges Clemente faced was racial prejudice. Many baseball fans, reporters, and players were rude or nasty to Clemente because he was black and Latino. However, he always defended his rights and the rights of others. Clemente said, "My greatest satisfaction comes from helping to erase the old opinion about Latin Americans and blacks."

On December 31, 1972, Clemente died in a plane crash only a few miles from where he was born. He was on his way to deliver aid to earthquake victims in Nicaragua. He was only thirty-eight, but he had become a baseball legend.

(1) Complete the main ideas for each paragraph in the chart below.

Paragraph	Main idea
First paragraph	Introduction to Roberto Clemente.
Second paragraph	The start of Clemente's baseball career.
Third paragraph	Clemente set baseball records and delighted fans.
Fourth paragraph	Clemente faced racial prejudice.
Fifth paragraph	Clemente died young but had become a baseball legend.

(2) What is the main idea of the whole passage? Put a check (✓) next to the correct idea below.
( ) Over eighteen seasons, Clemente set impressive records.
( ) Clemente helped many people and fought racial prejudice.
(✓) Clemente overcame many challenges, became a baseball legend and helped people.

DAY 24, pages 47 & 48

1 Read the word problem and write the number sentence below. Then answer the question.

(1) Selena wrapped presents for her friends. She used 1.6 yards of ribbon out of the 2.3 yards of ribbon she had. How much ribbon was left?

2.3 − 1.6 = 0.7

Ans. 0.7 yard of ribbon

(2) My mother had 3 pounds of flour. She used 0.3 pound to bake a cake. How much flour is left?

3 − 0.3 = 2.7

Ans. 2.7 pounds of flour

(3) We have 2.4 liters of orange juice in the fridge. We also have apple juice, but 500 milliliters less. How much apple juice do we have in the fridge?

500 mL = 0.5 L
2.4 − 0.5 = 1.9

Ans. 1.9 liters of apple juice

(4) Brad and Mark both have sticks. Brad's is 2.1 meters long. If Brad's stick is 50 centimeters longer than Mark's, how long is Mark's stick?

50 cm = 0.5 m
2.1 − 0.5 = 1.6

Ans. 1.6 meters

(5) My bag of pears is 0.8 kilograms heavier than your bag of pineapples. If my bag of pears weighs 2 kilograms, how much does your bag of pineapples weigh?

2 − 0.8 = 1.2

Ans. 1.2 kilograms

1 Read the excerpt from *The Tale of Mr. Tod* by Beatrix Potter. Then answer the questions below using words from the passage.

I have made many books about well-behaved people. Now, for a change, I am going to make a story about two disagreeable people, called Tommy Brock and Mr. Tod. Nobody could call Mr. Tod "nice." The rabbits could not bear him; they could smell him half a mile off. He was of a wandering habit and he had foxey whiskers; they never knew where he would be next.

One day he was living in a stick-house in the coppice, causing terror to the family of old Mr. Benjamin Bouncer. Next day he moved into a pollard willow near the lake, frightening the wild ducks and the water rats.

In winter and early spring he might generally be found in an earth amongst the rocks at the top of Bull Banks, under Oatmeal Crag.

He had half a dozen houses, but he was seldom at home.

The houses were not always empty when Mr. Tod moved out; because sometimes Tommy Brock moved in; (without asking leave).

Tommy Brock was a short, bristly, fat, waddling person with a grin; he grinned all over his face. He was not nice in his habits. He ate wasp nests and frogs and worms; and he waddled about by moonlight, digging things up.

His clothes were very dirty; and as he slept in the daytime, he always went to bed in his boots.

(1) What type of characters will the author write a story about?
The author will write about two disagreeable **people.**

(2) Put a check (✓) next to the words that describe Mr. Tod.
( ) nice ( ) unmoving (✓) smelly
(✓) wanderer (✓) never home ( ) musical
( ) animal-friendly (✓) unpredictable ( ) funny

(3) Put a check (✓) next to the words that describe Tommy Brock.
(✓) smiley ( ) tall (✓) short
( ) nice (✓) dirty ( ) talented
( ) normal (✓) strange ( ) smart

(4) Despite it being daytime, what did Tommy Brock do?
Tommy Brock slept **in the daytime.**

DAY 25, pages 49 & 50

❶ Read the word problem and write the number sentence below. Then answer the question.

(1) At the supermarket, my mother bought some meat for $14 and some vegetables for $12. If she paid with a $50-bill, how much change did she get? Using parentheses, write this down in a formula and then solve it.

$$50 - (14 + 12) = 24$$

Ans. $ 24

(2) Rina's book about animals has 256 pages. She read 64 pages yesterday and 57 pages today. How many pages are left?

$$256 - (64 + 57) = 135$$

Ans. 135 pages

(3) Jack saw a jacket that he wanted to buy. He bargained with the store owner, who discounted the price $5. The jacket was originally $78, and Jack paid with a $100 bill. How much change will Jack get?

$$100 - (78 - 5) = 27$$

Ans. $ 27

(4) Because a dress was not popular, the shopkeeper discounted it $12. It used to cost $68. Helen bought it and paid with a $100 bill. How much change did she get?

$$100 - (68 - 12) = 44$$

Ans. $ 44

(5) You went shopping and bought a sandwich for $7 and a book for $12. If you paid with a $20 bill, how much change did you get?

$$20 - (7 + 12) = 1$$

Ans. $ 1

❶ Read the passage. Then answer the questions below.

There was once a cook called Grethel, who wore shoes with red heels, and when she went out in them she gave herself great airs, and thought herself very fine indeed. When she came home again, she would take a drink of wine to refresh herself, and as that gave her an appetite. The master used to see that the table was properly laid, and, taking the great carving knife with which he meant to cut the fowls, he sharpened it upon the step.

Now it happened that one day her master said to her—

"Grethel, I expect a guest this evening; you must make ready a pair of fowls."

"Certainly, sir, I will," answered Grethel. So she killed the fowls, cleaned them, and plucked them, and put them on the spit, and then, as evening drew near, placed them before the fire to roast. And they began to be brown, and were nearly done, but the guest had not come.

Then Grethel called out to her master, "If the guest does not come, I must take the fowls away from the fire, but it will be a thousand pities if they are not eaten the moment they are at their best."

The master said, "I will run myself, and fetch the guest." As soon as the master had turned his back, Grethel laid the spit with the fowls on one side, and thought, "Standing so long by the fire there, makes one hot and thirsty; who knows when they will come? Meanwhile, I will run into the cellar and take a drink." So down she ran, took a jug, said, "Here's to me, Grethel," and took a good drink, and thought that wine should flow on, and that it was not good to have it interrupted, and so took yet another hearty draught.

(1) Put a check (✓) next to the words that describe Grethel.
(✓) angry   ( ) greedy   (✓) vain
(✓) cook   ( ) modest   ( ) dirty
( ) generous   ( ) stern   ( ) curious

And now they began to smell so good that Grethel saying, "I must find out whether they really are all right," licked her fingers, and then cried, "Well, I never! The fowls are good; it's a sin and a shame that no one is here to eat them!"

So she ran to the window to see if her master and his guest were coming, but as she could see nobody she went back to her fowls. "Why, one of the wings is burning!" she cried presently, "I had better eat it and get it out of the way." So she cut it off and ate it up, and it tasted good, and then she thought, "I had better cut off the other one, in case the master should miss anything." And when both wings had been disposed of she went and looked for the master, but still he did not come. "Who knows," said she, "whether they are coming or not? They may have put up at an inn." And after a pause she said again, "Come, I may as well make myself happy, and first I will make sure of a good drink and then of a good meal."

(2) Why does Grethel say she must eat the first wing?
Grethel must eat the first wing because it is _burning_.

(3) Why does Grethel say she should eat the whole meal?
Grethel says she should eat the whole meal because she may as well make herself _happy_.

DAY 26, pages 51 & 52

❶ Read the word problem and write the number sentence below. Then answer the question.

(1) Allison wants to have 4 candies for everyone at her party. There are 11 boys and 13 girls at her party. How many candies will she need? Remember to use a formula.

$$(11 + 13) \times 4 = 96$$

Ans. 96 candies

(2) Jack is buying food for his chicken and chicks. Each day, the chicks eat 30 seeds and the chicken eats 60 seeds. If Jack wants to feed them for a week, how many seeds must he buy?

$$(30 + 60) \times 7 = 630$$

Ans. 630 seeds

(3) Olive found a brush and paint set that she liked. The brush cost $5 and the paint cost only $3. How many sets can she buy if she has $96?

$$96 \div (5 + 3) = 12$$

Ans. 12 sets

(4) Gayle and her 2 brothers gathered their money and bought a $43 racing video game and a $44 basketball video game. What did each person pay?

$$(43 + 44) \div 3 = 29$$

Ans. $ 29

(5) In the pantry, the 2 cans of beans weigh 350 grams each, and the 4 cans of beets weigh 430 grams each. What is the total weight of the cans?

$$(2 \times 350) + (4 \times 430) = 2,420$$

Ans. 2,420 grams

❶ Read the passage. Then identify the statements as T (true) or F (false) according to the passage.

Just as she was in the middle of it her master came back. "Make haste, Grethel," he cried, "the guest is coming directly!" "Very well, master," she answered, "it will soon be ready." The master went to see that the table was properly laid, and, taking the great carving knife with which he meant to cut the fowls, he sharpened it upon the step.

Presently came the guest, knocking very genteelly and softly at the front door. Grethel ran and looked to see who it was, and when she caught sight of the guest she put her finger on her lip saying, "Hush! Make the best haste you can out of this, for if my master catches you, it will be bad for you; he asked you to come to supper, but he really means to cut off your ears! Just listen how he is sharpening his knife!"

The guest, hearing the noise of the sharpening, made off as fast as he could go. And Grethel ran screaming to her master. "A pretty guest you have asked to the house!" she cried. "How so, Grethel? What do you mean?" he asked.

"What indeed!" she said. "Why, he has gone and run away with my pair of fowls that I had just dished up."

"That's a pretty sort of conduct!" said the master, feeling very sorry about the fowls. "He might at least have left me one, that I might have had something to eat." And he called out to him to stop, but the guest made as if he did not hear him. Then he ran after him, the knife still in his hand, crying out, "Only one! Only one!" meaning that the guest should let him have one of the fowls and not take both. But the guest thought he meant to have only one of his ears, and he ran so much the faster that he might get home with both of them safe.

(1) The master sharpened his knife to cut the fowl.   **T**   F

(2) Grethel told the guest the truth.   T   **F**

(3) The guest believed Grethel.   **T**   F

(4) The guest was lying to escape about her master.   T   **F**

(5) Grethel told her master a lie.   **T**   F

(6) Grethel found a clever way to cover up the eaten meal.   **T**   F

DAY 27, pages 53 & 54

❶ Read the word problem and write the number sentence below. Then answer the question.

(1) Kate lost her dog. She made 250 flyers and gave 5 flyers each to 34 people. How many flyers did she have left over?

$$250 - 5 \times 34 = 80$$

Ans. 80 flyers

(2) Andy saved up $37. Today his aunt gave him $50 and told him to divide it with his sister evenly. How much money does Andy have now?

$$37 + 50 \div 2 = 62$$

Ans. $ 62

(3) A store has 1,000 gumballs. 820 gumballs weighs 2 kilograms. If you buy 1 kilogram of gumballs, how many gumballs will be left?

$$1,000 - 820 \div 2 = 590$$

Ans. 590 gumballs

(4) A store has 200 oranges divided equally in 5 crates. Tina bought 1 crate of oranges. Tina also bought 3 crates of apples. Each crate holds 30 apples. How many pieces of fruit did Tina buy?

$$(30 \times 3) + (200 \div 5) = 130$$

Ans. 130 pieces

(5) Dana divided 72 stickers into 24 gift bags equally. Barry took 4 bags home. How many stickers did Barry take home?

$$72 \div 24 \times 4 = 12$$

Ans. 12 stickers

❶ Read the excerpt from Black Beauty by Anna Sewell. Then answer the questions below.

One night, a few days after James had left, I had eaten my hay and was lying down in my straw fast asleep, when I was suddenly roused by the stable bell ringing very loud. I heard the door of John's house open, and his feet running up to the hall. He was back again in no time; he unlocked the stable door, and came in, calling out, "Wake up, Beauty! You must go well now, if ever you did," and almost before I could think he had got the saddle on my back and the bridle on my head. He just ran round for his coat, and then took me at a quick trot up to the hall door. The squire stood there, with a lamp in his hand.

"Now, John," he said, "ride for your life—that is, for your mistress' life; there is not a moment to lose. Give this note to Dr. White; give your horse a rest at the inn, and be back as soon as you can."

John said, "Yes, sir," and was on my back in a minute.

The gardener who lived at the lodge had heard the bell ring, and was ready with the gate open, and away we went through the park, and through the village, and down the hill till we came to the toll-gate. John called very loud and thumped upon the door; the man was soon out and flung open the gate.

"Now," said John, "do keep the gate open for the doctor; here's the money," and off he went again.

(1) What was Black Beauty doing when the stable bell rang?
Black Beauty was _lying_ down and _asleep_ in his straw.

(2) What did John do to get Black Beauty ready to ride?
John _got_ the _saddle_ on his back and the _bridle_ on his head.

(3) What did the squire tell John he must do?
The squire told John to _give_ a note to Dr. White.

(4) Describe the route that John and Black Beauty took to get to the toll-gate.
John and Black Beauty went _through_ the park, and the village, and _down_ the hill.

(5) How did John call for the man at the toll-gate?
John called very _loud_ for the man.

(6) Put a check (✓) next to the words that describe the scene.
(✓) alarming   ( ) lazy   ( ) humorous
( ) relaxed   (✓) important   ( ) sad
(✓) urgent   (✓) tense   ( ) boring

DAY 28, pages 55 & 56

❶ Read the word problem and write the number sentence below. Then answer the question.

(1) The teacher brought 21 dozen colored pencils to art class today. If she divided the pencils equally among 28 students, how many pencils did each student get?

$$21 \times 12 \div 28 = 9$$

Ans. 9 pencils

(2) Glen bought 6 packs of 45 stickers and 9 packs of 70 stickers. How many stickers did he get in all?

$$(45 \times 6) + (70 \times 9) = 900$$

Ans. 900 stickers

(3) Eddie bought 5 big bags of chips for $15. His brother bought 3 small bags and paid $6. How much more expensive were Eddie's bags of chips?

$$(15 \div 5) - (6 \div 3) = 1$$

Ans. $ 1

(4) My piggy bank has 37 coins in it. They are all pennies and nickels. If there are 5 more pennies than nickels, how many nickels do I have?

$$(37 - 5) \div 2 = 16$$

Ans. 16 nickels

(5) Shannon has 84 stamps and her sister has 50. How many stamps does Shannon have to give her sister so that they have the same amount?

$$(84 + 50) \div 2 = 67$$
$$84 - 67 = 17$$

Ans. 17 stamps

❶ Read the passage. Then answer the questions below.

There was before us a long piece of level road by the river side, John said to me, "Now, Beauty, do your best," and so I did. I wanted no whip nor spur, and for two miles I galloped as fast as I could lay my feet to the ground. I don't believe that my old grandfather, who won the race at Newmarket, could have gone faster. When we came to the bridge John pulled me up a little and patted my neck. "Well done, Beauty! Good old fellow," he said.

He would have let me go slower, but my spirit was up, and I was off again as fast as before. The air was frosty, the moon was bright; it was very pleasant. We came through a village, then through a dark wood, then downhill, till after eight miles' run we came to the town, through the streets and into the marketplace. It was all quite still except the clatter of my feet on the stones—everybody was asleep. The church clock struck three as we drew up at Dr. White's door.

(1) What did Black Beauty do for two miles?
Black Beauty _galloped_ as fast as he could for two miles.

(2) When they arrived at the bridge, what did John do?
John _pulled_ Black Beauty up and _patted_ his neck.

(3) Put a check (✓) next to the word that describes Black Beauty's run.
( ) relaxed   ( ) sluggish   (✓) fast

(4) Describe the route that John and Black Beauty took after the village.
John and Black Beauty went through a _dark_ wood, then _uphill_, then _downhill_.

(5) Write a **B** next to the sentence below that is a description only.
(D) There was a long piece of level road by the river side.
( ) John thumped upon the door.
(B) I galloped as fast as I could for two miles.
(D) The moon was bright.
( ) I was off again as fast as before.
(D) It was very pleasant.
(D) It was all quite still except the clatter of my feet on the stones.
(D) The air was frosty.

DAY 29, pages 57 & 58

❶ Read the word problem and write the number sentence below. Then answer the question.

(1) Jessica's mother weighs 54 kilograms, and that is twice as much as Jessica weighs. Jessica weighs 3 times as much as her baby sister. How much does Jessica's baby sister weigh?

$$54 \div 2 = 27$$
$$27 \div 3 = 9$$

Ans. 9 kilograms

(2) There are apples and melons in the fruit basket in the cafeteria. Altogether there are 35 pieces of fruit. If there are 4 times as many apples as melons, how many of each kind of fruit is in the basket?

$$35 \div (4 + 1) = 7$$
$$7 \times 4 = 28$$

Ans. Apples 28   Melons 7

(3) At Farmer William's farm, he has cows and horses. He has 3 times as many horses as cows. If there are 34 more horses than cows, how many of each does Farmer William have?

$$34 \div (3 - 1) = 17$$
$$17 \times 3 = 51$$

Ans. Cows 17   Horses 51

(4) Ted has 5 pieces of 8-inch tape, but he connects them so 2 inches overlap. How long is his new piece?

$$(8 \times 5) - (2 \times 4) = 32$$

Ans. 32 inches

(5) Sally's living room is 3 meters and 50 centimeters wide. Her pictures are 35 centimeters wide, and she wants the pictures spaced equally as shown below. How much space should she put between the pictures in her living room?

$$(350 - 35 \times 6) \div 7 = 20$$

Ans. 20 centimeters

❶ Read the passage. Then answer the questions below.

John rang the bell twice, and then knocked at the door like thunder. A window was thrown up, and Dr. White, in his nightcap, put his head out and said, "What do you want?"

"Mrs. Gordon is very ill, sir; master wants you to go at once; he thinks she will die if you cannot get there. Here is a note."

"Wait," he said, "I will come."

He shut the window, and was soon at the door.

"The worst of it is," he said, "that my horse has been out all day and is quite done up; my son had just been sent for, and he has taken the other. What is to be done? Can I have your horse?"

"He has come at a gallop nearly all the way, sir, and I was to give him a rest here; but I think my master would not be against it, if you think fit, sir."

"All right," he said; "I will soon be ready."

John stood by me and stroked my neck; I was very hot. The doctor came out with his riding-whip.

"You need not take that, sir," said John; "Black Beauty will go till he drops. Take care of him, sir, if you can; I should not like any harm to come to him."

"No, no, John," said the doctor, "I hope not," and in a minute we had left John far behind.

(1) What did John do to get the doctor to wake up?
John _rang_ the bell twice and _knocked_ at the door.

(2) Describe how John knocked at the door.
John knocked at the door like _thunder_.

(3) While they waited for the doctor, what did John do?
John _stroked_ Black Beauty's neck.

(4) Write an **A** next to the sentence below that is action only.
( ) John rang the bell twice.
( ) Dr. White's nightcap was droopy.
(A) Dr. White shut the window, and was soon at the door.
(A) John stood by me and stroked my neck.
(A) John knocked at the door.
( ) The knock was like thunder.
( ) I was very hot.
(A) In a minute we had left John far behind.

Way to go!

DAY 30, pages 59 & 60

❶ The table and the graph pictured here both show the temperature over one day. Answer the questions about the graph below.

Time (o'clock)	6	7	8	9	10	11	12	1	2	3	4	5	6
Temperature (C)	12	14	15	16	18	20	22	24	23	20	19	17	

(1) Write the appropriate label for the horizontal axis.
(2) Write the appropriate label for the vertical axis.
(3) Complete the line graph by placing each point and then connecting them with a line.
(4) Write the title in box A.

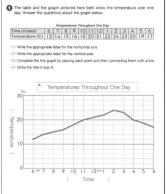

A   Temperatures Throughout One Day

❶ Read the passage. Then answer the questions below.

I will not tell about our way back. The doctor was a heavier man than John and not so good a rider; however, I did my very best. The man at the toll-gate had it open. When we came to the hill the doctor drew me up. "Now, my good fellow," he said, "take some breath." I was glad he did, for I was nearly spent, but that breathing helped me on, and soon we were in the park. Joe was at the lodge gate; my master was at the hall door, for he had heard us coming. He spoke not a word; the doctor went into the house with him, and Joe led me to the stable. I was glad to get home; my legs shook under me, and I could only stand and pant. I had not a dry hair on my body, the water ran down my legs, and I steamed all over. Joe used to say, like a pot on the fire. Poor Joe! He was young and small, and as yet he knew very little, and his father, who would have helped him, had been sent to the next village; but I am sure he did the very best he knew. He rubbed my legs and my chest, but he did not put my warm cloth on me, he thought I was so hot I should not like it. Then he gave me a pailful of water to drink; it was cold and very good, and I drank it all; then he gave me some hay and some corn, and thinking he had done right, he went away. Soon I began to shake and tremble, and turned deadly cold; my legs ached, my loins ached, and my chest ached, and I felt sore all over. Oh! how I wished for my warm, thick cloth, as I stood and trembled. I wished for John, but he had eight miles to walk, so I lay down in my straw and tried to go to sleep.

(1) How was the doctor different from John?
The doctor was a _heavier_ man than John and not as _good_ a rider.

(2) When they came to the hill, what did the doctor make Black Beauty do?
When they came to the hill, the doctor made Black Beauty take a _breath_.

(3) Describe Joe.
Joe was _young_ and _small_ and knew _little_ about caring for horses.

(4) Write an **A** next to the sentences below that are actions only.
( ) It was like a pot on fire.
(A) I began to shake and tremble.
(A) I lay down in my straw.

(5) Write a **D** next to the sentences below that are descriptions only.
(D) I was glad to get home.
(D) I had not a dry hair on my body.
( ) He rubbed my legs and my chest.

DAY 31, pages 61 & 62

Don't forget!
The volume of a cube that has 1-inch sides is 1 cubic inch and is written 1 in.³

❶ The following shapes were made by cubes with 1-inch sides. Calculate the volume of each shape below.

(1) (1 in.³)
(2) (2 in.³)
(3) (2 in.³)
(4) (3 in.³)
(5) (4 in.³)

❷ Calculate the volume of the following rectangular prisms. Answer in cubic inches.

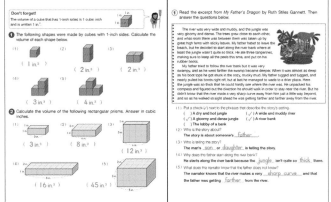

(1) (3 in.³)
(2) (8 in.³)
(3) (12 in.³)
(4) (16 in.³)
(5) (45 in.³)

❶ Read the excerpt from My Father's Dragon by Ruth Stiles Gannett. Then answer the questions below.

The river was very wide and muddy, and the jungle was very gloomy and dense. The trees grew close to each other, and what room there was between them was taken up by great high ferns with sticky leaves. My father hated to leave the beach, but he decided to start along the river bank where at least the jungle wasn't quite so thick. He ate three tangerines, making sure to keep all the peels this time, and put on his rubber boots.

My father tried to follow the river bank but it was very swampy, and as he went farther the swamp became deeper. When it was almost as deep as his boot tops he got stuck in the ooze, mucky muck. My father tugged and tugged, and nearly pulled his boots right off, but at last he managed to wade to a drier place. Here the jungle was so thick that he could hardly see where the river was. He unpacked his compass and figured out the direction he should walk in order to stay near the river. But he didn't know that the river made a very sharp curve away from him just a little way beyond, and so as he walked straight ahead he was getting farther and farther away from the river.

(1) Put a check (✓) next to the phrases that describe the story's setting.
( ) A dry and hot jungle   (✓) A wide and muddy river
(✓) A gloomy and dense jungle   (✓) A river bank
( ) The lobby of a bank

(2) Who is the story about?
The story is about someone's _father_.

(3) Who is telling the story?
The man's _son_ or daughter is telling the story.

(4) Why does the narrator start along the river bank?
He starts along the river bank because the _jungle_ isn't quite so thick there.

(5) What does the narrator know that his father does not know?
The narrator knows that the river makes a very _sharp curve_ and that the father gets _farther_ from the river.

DAY 32, pages 63 & 64

Don't forget!
The volume of a cube that has 1-centimeter sides is 1 cubic centimeter and a written 1 cm³.

❶ The following shapes were made by cubes with 1-cm sides. Calculate the volume of each shape below.

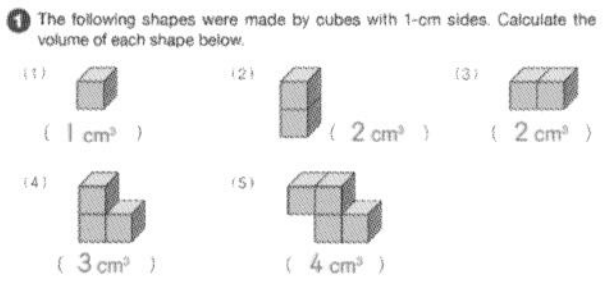

(1) (1 cm³)
(2) (2 cm³)
(3) (2 cm³)
(4) (3 cm³)
(5) (4 cm³)

❷ Calculate the volume of the following rectangular prisms. Answer in cubic centimeters.

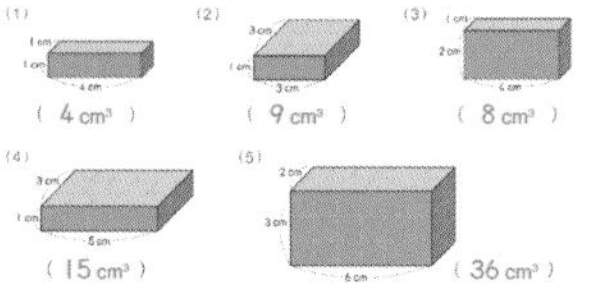

(1) (4 cm³)
(2) (9 cm³)
(3) (8 cm³)
(4) (15 cm³)
(5) (36 cm³)

❶ Read the passage. Then answer the questions below.

It was very hard to walk in the jungle. The sticky leaves of the ferns caught at my father's hair, and he kept tripping over roots and rotten logs. Sometimes the trees were clumped so closely together that he couldn't squeeze between them and had to walk a long way around.

He began to hear whispery noises, but he couldn't see any animals anywhere. The deeper into the jungle he went the sure he was that something was following him, and then he thought he heard whispery noises on both sides of him as well as behind. He tried to run, but he tripped over more roots and, and the noises only came nearer. Once or twice he thought he heard something laughing at him.

At last he came out into a clearing and ran right into the middle of a lot that he could see anything that might try to attack him. Was he surprised when he looked and saw fourteen green eyes coming out of the jungle all around the clearing, and then the green eyes turned into seven tigers! The tigers walked around him in a big circle, looking hungrier all the time, and then they sat down and began to talk.

"I suppose you thought we didn't know you were trespassing in our jungle!"

(1) Why was it hard for the father to walk in the jungle?
It was hard for the father to walk in the jungle because he kept tripping over _roots_ and _rotten logs_.

(2) Complete the chart below that shows the cause of each of the father's actions.

Cause	Effect
The trees were clumped close together.	The father had to _walk_ a long way around.
The father heard whispery noises all around him.	He tried to _run_.
The father wanted to see anything that might try to attack him.	He _ran_ into the middle of a _clearing_.

(3) What did the tigers think the father was doing?
The tigers thought that the father was _trespassing_ in their jungle.

DAY 33, pages 65 & 66

1 Convert the measurements below.

(1) 1 pt. = **2** cups
(2) 2 pt. = **4** cups
(3) 3 pt. = **6** cups
(4) 2 cups = **1** pt
(5) 1 cup = **0.5** pt.
(6) 4 cups = **2** pt.

2 Convert the measurements below.

(1) 1 cup = **8** fl. oz.
(2) 2 cups = **16** fl. oz.
(3) 4 cups = **32** fl. oz.
(4) 8 fl. oz. = **1** cup
(5) 24 fl. oz. = **3** cups
(6) 40 fl. oz. = **5** cups

3 How much water is in each measuring cup? Answer in two different units.

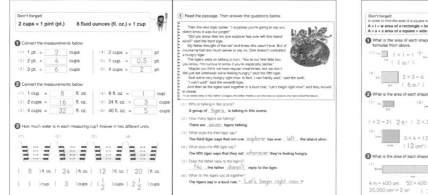

(1) (**8**) fl. oz. (**1**) cup
(2) (**24**) fl. oz. (**3**) cups
(3) (**12**) fl. oz. (**1½**) cups
(4) (**20**) fl. oz. (**2½**) cups

1 Read the passage. Then answer the questions below.

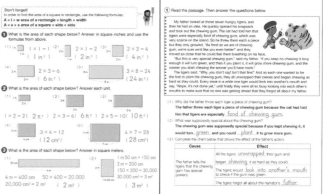

Then the next tiger spoke. "I suppose you're going to say you didn't know that our jungle?" ...

(1) Who is talking in this scene?
A group of **tigers** is talking in this scene.
(2) How many tigers are talking?
There are **seven** tigers talking.
(3) What does the third tiger say?
The third tiger says that not one **explorer** has ever **left** the island alive.
(4) What does the fifth tiger say?
The fifth tiger says that they eat **whenever** they're feeling hungry.
(5) Does the father reply to the tigers?
No , the father **doesn't** reply to the tiger.
(6) What do the tigers say all together?
The tigers say in a loud roar, " **Let's begin right now** "

DAY 34, pages 67 & 68

1 What is the area of each shape below? Answer in square inches and use the formulas from above.

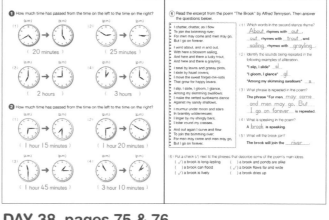

(1) 1 × 1 = **1** (**1** in.²)
(2) 2 × 1 = **2** (**2** in.²)
(3) 2 × 2 = **4** (**4** in.²)
(4) 2 × 3 = **6** (**6** in.²)
(5) 3 × 8 = **24** (**24** in.²)

2 What is the area of each shape below? Answer each unit.

(1) 1 × 2 = **2** (**2** ft.²)
(2) 2 × 3 = **6** (**6** ft.²)
(3) 2 × 5 = **10** (**10** ft.²)
(4) 3 × 4 = **12** (**12** cm²)
(5) 4 × 7 = **28** (**28** cm²)

3 What is the area of each shape below? Answer in square meters.

4 m = 400 cm   50 × 400 = 20,000
20,000 cm² = 2 m²   (**2** m²)

1 m 50 cm = 150 cm
150 × 200 = 30,000
30,000 cm² = 3 m²   (**3** m²)

1 Read the passage. Then answer the questions below.

My father looked at those seven hungry tigers, and then he had an idea. ...

(1) Why did the father throw each tiger a piece of chewing gum?
The father threw each tiger a piece of chewing gum because the cat had told him that tigers are especially **fond** of chewing gum.
(2) What was supposedly special about this chewing gum?
The chewing gum was supposedly special because if you kept chewing it, it would turn **green** , and you could **plant** it to grow more gum.
(3) Complete the chart below that shows the effect of the father's action.

Cause	Effect
The father tells the tigers that the chewing gum has special powers	All the tigers **unwrapped** their gum and began **chewing** it as hard as they could.
	The tigers would **look into another's mouth** to check if the gum was green.
	The tigers forgot all about the narrator's **father**.

DAY 35, pages 69 & 70

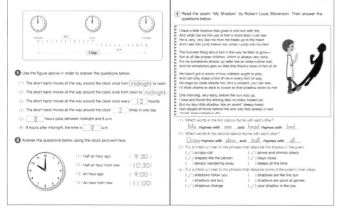

1 Use the figure above in order to answer the questions below.

(1) The short hand moves all the way around the clock through the center is called the **diameter**... from **midnight** to noon.
(2) The short hand moves all the way around the clock once from noon to **midnight**.
(3) The short hand moves all the way around the clock every **12** hours.
(4) The short hand moves all the way around the clock **2** times in one day.
(5) **8** hours pass between midnight and 8 a.m.
(6) 9 hours after midnight, the time is **9** a.m.

2 Answer the questions below using the clock pictured here.

(1) Half an hour ago (9:30)
(2) Half an hour from now (10:30)
(3) An hour ago (9:00)
(4) An hour from now (11:00)

1 Read the poem "My Shadow" by Robert Louis Stevenson. Then answer the questions below.

I have a little shadow that goes in and out with me,
And what can be the use of him is more than I can see. ...

(1) Which words in the first stanza rhyme with each other?
Me rhymes with **see** , and **head** rhymes with **bed** .
(2) Which words in the second stanza rhyme with each other?
Grow rhymes with **slow** , and **ball** rhymes with **all** .
(3) Put a check (✓) next to the phrases that describe the shadow in the poem.
(✓) a copy-cat   ( ) grows and shrinks slowly
(✓) shaped like the person   ( ) stays close
( ) always wandering away   ( ) sleeps all the time
(4) Put a check (✓) next to the phrases that describe some of the poem's main ideas.
(✓) shadows follow you   ( ) shadows are like the sun
( ) shadows are lazy   ( ) shadows are good at games
(✓) shadows change   (✓) your shadow is like you

DAY 36, pages 71 & 72

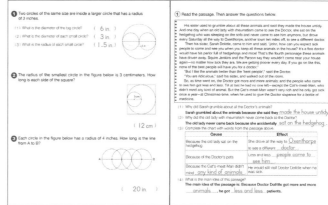

1 How much time has passed from the time on the left to the time on the right?

(1) (20 minutes)
(2) (25 minutes)
(3) (2 hours)
(4) (3 hours)

2 How much time has passed from the time on the left to the time on the right?

(1) (1 hour 15 minutes)
(2) (1 hour 20 minutes)
(3) (1 hour 45 minutes)
(4) (3 hour 10 minutes)

1 Read the excerpt from the poem "The Brook" by Alfred Tennyson. Then answer the questions below.

I chatter, chatter, as I flow
To join the brimming river, ...

(1) Which words in the second stanza rhyme?
About rhymes with **out** , **out** rhymes with **trout** , and **sailing** rhymes with **grayling** .
(2) Identify the sounds being repeated in the following examples of alliteration.
"I slip, I slide" **sl**
"I gloom, I glance" **gl**
"Among my skimming swallows" **s**
(3) What phrase is repeated in the poem?
The phrase "For men **may come and men may go, But** I go on forever" is repeated.
(4) What is speaking in the poem?
A **brook** is speaking.
(5) What will the brook join?
The brook will join the **river** .
(6) Put a check (✓) next to the phrases that describe some of the poem's main ideas.
(✓) a brook is long-lasting ( ) a brook and ponds are alike
( ) a brook can flood   (✓) a brook flows far and wide
(✓) a brook is lively   ( ) a brook dries up

DAY 37, pages 73 & 74

1 Write the appropriate number in each box below.

(1) If the radius of a circle is 2 centimeters, the diameter is **4** centimeters.
(2) If the diameter of a circle is 2 centimeters, the radius is **1** centimeter(s).
(3) If the radius of a circle is 6 centimeters, the diameter is **12** centimeters.
(4) If the diameter of a circle is 6 centimeters, the radius is **3** centimeters.

2 Use a compass to draw a circle in each box below.

(1) Draw a circle with a radius of 2 centimeters.
(2) Draw a circle with a diameter of 5 centimeters.

Please ask your parents to check if your answer is correct.

1 Read the excerpt from The Story of Doctor Dolittle by Hugh Lofting. Then answer the questions below.

Once upon a time, many years ago when our grandfathers were little children—there was a doctor; and his name was Dolittle—John Dolittle, M.D. "M.D." means that he was a proper doctor and knew a whole lot. ...

(1) When does the story take place?
The story takes place many **years** ago.
(2) Who is the main character?
John Dolittle is the main character.
(3) Where is the setting of the story?
The setting of the story is the **house** of Dr. Dolittle at the edge of a little town called **Puddleby-on-the-Marsh** .
(4) What does Dr. Dolittle's sister, Sarah Dolittle, do?
Sarah Dolittle is Dr. Dolittle's **housekeeper** .
(5) Why does Dr. Dolittle keep so many pets?
Dr. Dolittle keeps so many pets because he is very **fond of animals** .
(6) Who are Dr. Dolittle's favorite pets?
Dr. Dolittle's favorite pets are **Dab-Dab** the **duck** , **Jip** the **dog** , **Gub-Gub** the **baby pig** , **Polynesia** the **parrot** and the **owl Too-Too** .
(7) Why does everyone think Dr. Dolittle is a clever man?
Everyone thinks Dr. Dolittle is a clever man because he is a **doctor** .

DAY 38, pages 75 & 76

1 Two circles of the same size are inside a larger circle that has a radius of 3 inches.

(1) What is the diameter of the big circle? (6 in.)
(2) What is the diameter of each small circle? (3 in.)
(3) What is the radius of each small circle? (1.5 in.)

2 The radius of the smallest circle in the figure below is 3 centimeters. How long is each side of the square?

(12 cm)

3 Each circle in the figure below has a radius of 4 inches. How long is the line from A to B?

(20 in.)

1 Read the passage. Then answer the questions below.

His sister used to grumble about all these animals and said they made the house untidy. And one day when an old lady with rheumatism came to see the Doctor, she sat on the hedgehog who was sleeping on the sofa and never came to see him anymore, but drove every Saturday all the way to Oxenthorpe, another town ten miles off, to see a different doctor. ...

(1) Why did Sarah grumble about all the Doctor's animals?
Sarah grumbled about the animals because she said they **made the house untidy.**
(2) Why did the old lady with rheumatism never come back to the Doctor?
The old lady never came back because she accidentally **sat on the hedgehog** .
(3) Complete the chart with words from the passage above.

Cause	Effect
Because the old lady sat on the hedgehog	She drove all the way to **Oxenthorpe** to see a different **doctor** .
Because of the Doctor's pets	Less and less **people come to see him** .
Because the Cat's-meat Man didn't need **any kind of animals** .	He would visit Doctor Dolittle when he was sick.

(4) What is the main idea of this passage?
The main idea of the passage is: Because Doctor Dolittle got more and more **animals** , he got **less and less** patients.

DAY 39, pages 77 & 78

1 Write the appropriate number in each box below.

(1) If the diameter of a sphere is 6 centimeters, the radius is **3** centimeters.
(2) If the diameter of a sphere is 5 centimeters, its radius is **2** centimeters and **5** millimeters.
(3) If the radius of a sphere is 4 centimeters, the diameter is **8** centimeters.
(4) If the radius of a sphere is 5 centimeters, the diameter is **10** centimeters.

2 As pictured on the right, you have a sphere inside a box.

(1) How long is each side of the box? (12 in.)
(2) How long is the diameter of the sphere? (12 in.)
(3) How long is the radius of the sphere? (6 in.)

3 As pictured on the right, you have six balls that fit snugly inside one box that is 15 inches wide.

(1) What is the diameter of each ball? (5 in.)
(2) What is the length of the box? (10 in.)

1 Read the passage. Then answer the questions below.

Sixpence a year wasn't enough to live on—even in those days, long ago; and if the Doctor hadn't had some money saved up in his money-box, no one knows what would have happened. ...

(1) Complete the chart with words from the passage above.

Cause	Effect
Sixpence a year wasn't enough to live on.	The Doctor spent his **saved** money.
He kept on getting still more pets.	The Doctor sold his **piano** .
The money from his piano wasn't enough.	The Doctor sold his **brown suit** and became **poorer and poorer** .

(2) Identify the statements as T (true) or F (false) according to the passage.
(1) The money was in the Doctor's bureau-drawer. T **F**
(2) Sixpence a year was just enough to live on. T **F**
(3) The Doctor kept losing his pets. T **F**
(3) Who didn't care that Doctor Dolittle didn't have any money?
The **dogs** and the **cats** and the **children** didn't care that Doctor Dolittle didn't have any money.

DAY 40, pages 79 & 80

1 Sort the triangles below into equilateral and isosceles triangles by putting each letter next to the correct category.

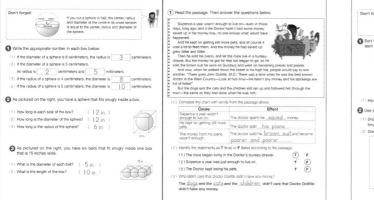

(1) equilateral triangle (a, d, f)
(2) isosceles triangle (b, e, j)

2 Use your ruler and compass to draw the triangles below.

(1) Draw a triangle with sides that are 4 centimeters, 2 centimeters, and 4 centimeters long.
(2) Draw a triangle with sides that are all 3 centimeters long.

Please ask your parents to check if your answer is correct.

1 Read the passage. Then answer the questions below.

It happened one day that the Doctor was sitting in his kitchen talking with the Cat's-meat-Man who had come to see him with a stomach-ache. ...

(1) Who is speaking in this passage?
The **Cat's-meat-Man** and the **Doctor** are speaking in this passage.
(2) Who is listening in this scene?
The **parrot(Polynesia)** is listening.
(3) What is the Cat's-meat-Man's suggestion?
The Cat's-meat-Man's idea is that the Doctor should become an **animal-doctor** .
(4) What are some supporting details for Cat's-meat-Man's suggestion?
(a) The Doctor knows all about **animals** .
(b) The Doctor is a wonderful **book** about **cats** .
(c) The Doctor can make a lot of **money** doctoring animals.
(5) What does the Doctor tell the Cat's-meat-Man not to do?
The Doctor tells him not to put something in the **meat** to make the animals **sick** .

DAY 41, pages 81 & 82

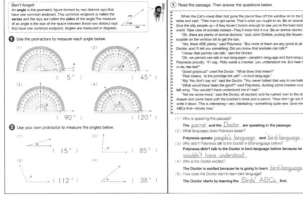

Don't forget!
An **angle** is the geometric figure formed by two distinct rays that have one common endpoint. This common endpoint is called the **vertex**, and the rays are called the **sides** of the angle. The measure of an angle is the size of the space between those two distinct rays that have one common endpoint. Angles are measured in degrees.

① Use the protractors to measure each angle below.

(5°)　(45°)

(90°)　(120°)

② Use your own protractor to measure the angles below.

(1) (15°)　(2) (85°)

(3) (112°)　(4) (38°)

① Read the passage. Then answer the questions below.

(1) Who is speaking in this passage?
The __parrot__ and the __Doctor__ are speaking in the passage.
(2) What language does Polynesia speak?
Polynesia speaks __people's language__ and __bird-language__.
(3) Why didn't Polynesia talk to the Doctor in bird-language before?
Polynesia didn't talk to the Doctor in bird-language before because he __wouldn't have understood__.
(4) Why is the Doctor excited?
The Doctor is excited because he is going to learn __bird-language__.
(5) How does the Doctor start to learn bird-language?
The Doctor starts by learning the __Birds' ABCs__ first.

DAY 42, pages 83 & 84

① Review the example. Then find the measure of angle A in each illustration below.

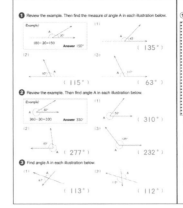

(Example) 180 − 30 = 150　Answer 150°

(1) (135°)
(2) (115°)　(3) (63°)

② Review the example. Then find angle A in each illustration below.

(Example) 3∠G − 30 = 330　Answer 330°

(1) (310°)
(2) (277°)　(3) (232°)

③ Find angle A in each illustration below.

(113°)　(112°)

① Read the passage. Then answer the questions below.

(1) What did Doctor Dolittle write down in his book?
The Doctor wrote down __bird words__ in his book.
(2) Who was talking to Doctor Dolittle with his nose?
The __dog (Jip)__ was talking to the Doctor with his nose.
(3) How was the Doctor able to learn the language of the animals?
The Doctor was able to learn the language of the animals with the help of the __parrot (Polynesia)__.
(4) Why don't animals always speak with their mouths?
Animals don't always speak with their mouths because sometimes they don't want to make __a noise__.
(5) Who brought their sick animals to Doctor Dolittle?
The __old ladies__ and the __farmers__ brought their sick animals.

DAY 43, pages 85 & 86

① Multiply or divide.

(1)
```
  4 0
×   2
─────
  8 0
```
(5)
```
  3 2
× 2 3
─────
  9 6
 6 4
─────
 7 3 6
```
(9) 55)6R45 → 55)375
(13) 29)5R27 → 29)1506

(2)
```
  2 7 3
×     3
───────
  8 1 9
```
(6)
```
  1 1 5
× 1 3 4
───────
```
(10) 8)5R46 → 8)5846
(14) 18)30R10 → 18)550

(3)
```
  3 0 7
×     7
───────
2 1 4 9
```
(7)
```
  2 3 0
× 1 2 5
───────
```
(11) 5)1R06 → 5)806
(15) 328)5R170 → 328)1810

(4)
```
  2 0 1
× 1 1 3
───────
```
(8)
```
  4 0
×   7
```
(12) 123)2R22 → 123)222
(16) 97)7R53 → 97)732

② Stamps are sold in rolls of 45. If Jane buys 12 rolls, how many stamps did she get?
45 × 12 = 540　**Ans.** 540 stamps

③ There are 196 roses, and the florist split them evenly into 14 bunches. How many roses are in each bunch?
196 ÷ 14 = 14　**Ans.** 14 roses

① Read the passage. Then answer the questions below.

(1) Choose words from the passage to complete the definitions below.
__coveted__ — wanted to have something
__prettier__ — more attractive
__assured__ — convinced; got rid of any doubts
__manipulation__ — changes
__whimper__ — whine; cry; sob
__diverted__ — distracted; sidetracked

(2) How did the big monkey trick the little monkey?
The big monkey convinced the little monkey that the cake would be __prettier__ if it were more like the __moon__.

(3) Complete the chart with words from the passage above.

Cause	Effect
The little monkey had a cake that the big monkey coveted.	The big monkey __made a plan to__ __get the cake.__
The big monkey took a big mouthful.	The little monkey began to __whimper__.
The big monkey __swallowed__ __the rest of__ the cake.	The big monkey's __stomach__ hurt.

DAY 44, pages 87 & 88

① Calculate.

(1)
```
  1.7
+ 2.1
─────
  3.8
```
(3)
```
 12.5
−  1.5
──────
 11.0
```
(5)
```
  0.5
+10.7
─────
 11.2
```
(7)
```
  1.07
+ 1.40
──────
  2.47
```

(2)
```
  2.1
− 0.4
─────
  1.7
```
(4)
```
   3.7
+ 11.5
──────
  15.2
```
(6)
```
 12.5
−  8.0
──────
  4.5
```
(8)
```
  2.30
− 0.68
──────
  1.62
```

② Calculate.

(1) $\frac{3}{5} + \frac{1}{5} = \frac{4}{5}$
(3) $\frac{4}{9} + \frac{5}{9} = 1$
(5) $\frac{6}{7} − \frac{3}{7} = \frac{3}{7}$

(2) $\frac{2}{7} + \frac{3}{7} = \frac{5}{7}$
(4) $\frac{4}{5} − \frac{1}{5} = \frac{3}{5}$
(6) $\frac{7}{9} − \frac{5}{9} = \frac{2}{9}$

③ Cindy's bag weighs 2.5 kilograms. Her sister's bag is 700 grams heavier. How much does her sister's bag weigh? Answer in kilograms.
2.5 + 0.7 = 3.2　**Ans.** 3.2 kilograms

④ Fred's living room is 3 meters and 45 centimeters wide. His pictures are 40 centimeters wide, and he wants the pictures spaced equally as shown below. How much space should he put between the pictures in his living room?

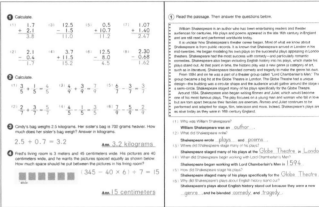

(345 − 40 × 6) ÷ 7 = 15
Ans. 15 centimeters

① Read the passage. Then answer the questions below.

(1) Who was William Shakespeare?
William Shakespeare was an __author__.
(2) What did Shakespeare write?
Shakespeare wrote __plays__ and __poems__.
(3) Where did Shakespeare stage many of his plays?
Shakespeare staged many of his plays at the __Globe Theatre__ in __London__.
(4) When did Shakespeare begin working with Lord Chamberlain's Men?
Shakespeare began working with Lord Chamberlain's Men in __1594__.
(5) How did Shakespeare stage his plays?
Shakespeare staged many of his plays specifically for the __Globe Theatre__.
(6) Why did Shakespeare's plays about English history stand out?
Shakespeare's plays about English history stood out because they were a new __genre__, and he blended __comedy__ and __tragedy__.

DAY 45, pages 89 & 90

① How much time has passed from the time on the left to the time on the right?

(1) (1 hour 50 minutes)
(2) (1 hour 35 minutes)

② As pictured on the right, you have a sphere that fits snugly inside a box.

(1) How long is each side of the box?　(10 in.)
(2) How long is the diameter of the sphere?　(10 in.)
(3) How long is the radius of the sphere?　(5 in.)

③ Find angle A in each illustration below.

(1) (117°)
(2) (108°)

④ Calculate the volume of the following rectangular prisms. Answer in cubic inches.

(1) (16 in.³)
(2) (30 in.³)

① Read the passage. Then answer the questions below.

(1) Describe the way the bucket and the bike must be moving in order to do the trick.
The bucket and the bike must be moving __fast enough__ in order to do the trick.
(2) What can the professional cyclist do because of centrifugal and centripetal force?
The professional cyclist can ride __upside-down__ in a loop.
(3) What is the main idea of the whole passage?
The main idea is that a cyclist takes advantage of __centrifugal__ force and __centripetal__ force to ride upside-down in a loop.
(4) What details support the main idea? Put a check (✓) next to the answers.
(✓) The cyclist and the bike continue to move in a straight line.
( ) The force of gravity holds objects down.
(✓) The track keeps the cyclist moving in a circular direction.
( ) Many inventions use centrifugal and centripetal force.